ANATOMY OF A YOUTHFUL WOMAN AND MAN

By

Linda S Sorrells

ISBN: 978-1-960321-04-6 (Paperback)

978-1-960321-05-3 (Hardcover)

Disclaimer

Caring Physicians seek a complete diagnosis of each patient and are familiar with the strengths of the medications that hormonal therapy prescribes. If, as you read this book, you feel it affords you an opportunity for self-diagnosis, it can be very hazardous. For example, a person with an impaired heart, liver, or kidney function can be at risk if she tries to self-medicate through hormones without correct medical supervision, which would weigh all medical factors. Even a relatively safe dose should not be taken without knowing that there are specific risks for certain individuals. All information should be carefully weighed by a knowledgeable physician familiar with hormone replacement before prescribing a list of therapeutic goals for an individual hormone replacement therapy program.

Dedication

This book is dedicated to my Parents, Joseph Lord, and Geneva Brooks.

Thank You for instilling in me a faith that can overcome every obstacle.

Acknowledgment

My entire experience with Writer Cosmos has been a success and remarkable during every touchpoint. My biggest dreams come true, which was to be a first-time published author. Harry was very helpful, along with Vincent sharing with me the best next steps. I appreciate their honesty and looking forward to my next book project.

About the Author

Linda S. Sorrells is a pastor of Agape Shalom way of ministries in California. She is a mother of four and has several grandchildren and great-grandchildren. Being a vocational nurse and medical assistant working with doctors and going through hormone authority adventures, I decided to become a Guinea pig. I chose to take the hormones, which has helped me remain young.

I'm 71 years old, and I plan on taking more hormones now that I am older. I did go against the doctor's orders by injecting myself with messaging, and I had to have Crystal removed from my right press for doing against what his orders were telling me. Still, I am continuing to further my self-education. I am an anti-trust lawyer, a medical assistant, and a pastor. As I said, I was diagnosed with clinical depression several years ago. Also, I like doing comedy acts and convincing others about the flaw in my ministry.

I have written this book which has taken 17 years of research. I have my own little doctor's kit from the Great Physician, and that's the Lord Jesus Christ. He is also my counselor. I'm basically headed for new horizons in the next book project.

CONTENTS

Foreword

Lying strapped to the rescue board, which, after six hours, was still covered with snow, I closed my mind to the continued questioning about my age. "This is really necessary for the medical report," the doctor urged. "We really need to know your age," he continued. Realizing that my chronological age would be a factor to be considered in determining the type of emergency treatment, I responded by saying that I had stopped telling my age after thirty-nine. Still, for the record, I would allow the doctor to write fifty in the necessary blank.

I might as well have said one hundred. There was a hush from both doctors and nurses, as their faces registered immediate surprise and indicated that I must be overreacting in order to get a rise out of them. In disbelief, one of the doctors stated that they had guessed my age between thirty to thirty-five at the most. Now, three months into the recovery period from a serious, broken hip from a mid-air twist while skiing, and now swimming one and one-half to two miles daily, driving my five-speed sports car, and working on at least five business deals daily, dating several times a week, laughing, joking, loving life — even on crutches with accompanying pain, I have come to realize that the majority of my female counterparts have not been or will not be so fortunate in life unless I assist my doctor in telling the medical secret which has helped me and allowed me to continue to be active, vibrant, and youthful the way I remember feeling in my early twenties.

To this end, and for your benefit, I said, "Please, doctor, you must write a book and dedicate it to all women who want to feel great, look young, and feel truly alive." How well I remember going through a period of depression about ten years ago when I went to the best doctors at a leading medical clinic.

I complained of back aches, feeling tired, dull, weary, and wanting to "throw in the towel," a feeling like I was slowly dying. After the most comprehensive examinations and fifteen hundred dollars in lab fees, the consulting doctors agreed that the only problem they could find was a low hormone count. Each doctor agreed that they did not believe in prescribing hormones, but in my case, where the hormone count has dropped nearly to zero, they might consider making an exception and prescribing the most minimal amount of .0325 per day for six months to a year. Perhaps in a year, they stated, I would be noticing some improvement. The dangers of cancer were reviewed for me again and again.

Six months later, I did begin feeling a twinge of improvement. At that time, I called my personal friend, Dr. Beals, and told him my story. He said to come in, and we'll give you injections twice a week, giving you the maximum dosages. Now, my warning – when my prescription for minimum amounts of estrogen was written – was that I should never allow any doctor to give me hormone injections for fear of inducing cancer. Trembling inside, I said, "You know, doctor, I have been warned so many times about cancer that I've been afraid to accept injections, but at this point in my life, I want to live life to the fullest. if I have to make a choice, I'll choose fewer years of healthy living instead of many years of depression, lack of energy, and feeling like I'm slowly dying." It was only a few days after receiving my first injection of hormones that I started feeling as if my starving and craving for something was finally being satisfied. The ten years that followed have been the best years of my life. To think that I could have been feeling like this in my thirties while raising my three children and family, going through a divorce, working two jobs, and pursuing my M.A. and administrative credentials all at the same time.

I believe that hormonal replacement restored my zest for life. A simple pap test focused on my physical needs ten years ago and fortunately, the author of this book was the pioneer who gave me a second chance to feel great again. You owe it to yourself, your family and loved ones, to understand your own hormonal needs. I urge you to have an open mind, and decide how you want to feel. Seek a doctor who will give you the hormones for which your body may be starving. You deserve to enjoy life as long as you live!

The Primetime of life can indeed be expanded, as explained in this book. Women can take charge of their destiny through this sexual revolution that will keep them younger and more energetic. Greater physical, emotional, and sexual self-confidence will enhance interactions with family members and business relationships. Each woman, now living, has a positive or tragic effect on her children, spouse, friends, and business acquaintances. And thus, she unwittingly colors the lives of hundreds of people!

This is a documented story about hormonal replacement therapy and how it can help women stay young, thus reducing chances of heart attack, stroke, cancer of the breast, uterus, and ovaries, depression, painful and often fatal fractures of osteoporosis, and premature aging.

A woman's sexual hormonal revolution reveals the miracle of how hormonal replacement therapy not only helps prevent many tragic illnesses, but premenstrual tension, menopause, and premature aging will also respond to hormonal replacement.

Introduction

This book was written to answer questions for millions of women caught in the confusion and misunderstanding of hormones, menstrual problems, menopause, arthritis, cancer, and aging. I intend to explain how the flame, fervor, and power a young woman enjoys at age twenty-five can be retained to age forty-five or even sixty-five.

Hormonal balance is a birthright of every woman. Failing to maintain a hormonal balance could cause a twenty-year drought in a woman's life spring. The depletion of inner hormonal resources is unnecessary, archaic, and possibly even a sexist problem within our medical profession.

Whereby there is little or no emphasis on hormone therapy in current research. Whether this is due to a lack of interest, inaccurate medical information, or just a medical controversy is debatable.

The author intends to share information that will be valuable in dispelling fears and myths for today's women to allow these women to take their destinies into their own hands. Any woman can learn how to obtain hormones that could be rightfully hers. Assistance will be outlined, which will assist women in securing a medical doctor who will help them in restoring and maintaining the vibrance and sexual aliveness that God intended for each woman and to which the poets allude to in their writings throughout history.

Armed with the knowledge contained in this book, a woman can become all things — the inspiration, drive, excitement, passion, and color to those around her, not only at twenty-five but also at fifty, seventy-five, and even one hundred.

To my medical cohorts who may have a client bring this book into your office demanding hormonal restoration, may I say that I feel sad regarding any uncomfortable encounter that may have been precipitated by this exposure? The need for this book could have been avoided if the medical establishments had listened to the findings of many excellent physicians who tried to inform doctors regarding the benefits of endocrine therapy and the tragedies that such therapy would have prevented during the past thirty-five years. The medical establishments could have told the story in balance, including all the benefits of endocrine therapy, and therefore, many more physicians would have understood the importance of hormonal restoration. Dr. Wilson, one of the pioneers, attempted as recently as 1967 to educate the entire population regarding hormonal therapy in the book Forever Feminine, which is still highly controversial and criticized within medical circles.

Chapter 1: The World's Greatest Untapped Natural Resource

Women, by nature, are often more spiritual and imaginative than men and, therefore, serve as their family's moral, religious, and spiritual motivational spirit. This trait can, but shouldn't have to, survive great adversity from the husband or the family's circumstances. It pays big dividends when the husband lends more significant support to the spiritual leadership of the family. That doesn't mean that the husband has to be religious or pious, but simply saying a prayer at dinner or lending strength to the family can set the mood for leadership when questions from the children arise later. A man owes it to himself to appreciate the qualities of strength in his woman and to both nurture and benefit from them. If he nurtures these qualities, they will be there in time of need. A successful couple truly enriches each other, aiding each other through the rough passages of life.

A woman is to a man, his home, and his family, as water is to a vine. The more the woman gives of herself, the stronger and more beautiful she grows. A vine grows until it covers an entire house so that when one views it, the eye is met with the greatness of strength and beauty of the vine and not just bare walls. If a woman is dried up in spirit, attitudes, or hormones – which are her life spring –she cannot give what she does not possess. If she possessed a strong spirit at one time in her life, but that spirit is now shrinking, then the leadership she gives to her family will be crippled.

A man is blessed to have a woman who has a great spirit with the sweetness of life with a reserve of life spring hormones. She is like a goddess of love, a friend, and a companion on this hazardous road of life with her husband. She will be a source of strength in times of need. This actually enables a man to be stronger, for he is satisfied and patient and will have fewer frustrations. In this blessed union, we see a man adoring and enjoying a loving relationship with his mate. Her essence improves his ability to perceive things clearly.

When people observe this couple, they are often pleasantly blessed, for they feel that they are in the presence of great harmony, peace, and strength. Competition, jealousy, and tension are not present. Their love is strong and deserving of trust.

My father once took me to an ancient orchard where great trees stood tall and bare as if their arms were praying to the heavens for life. He explained to me that when he was a boy, he used to go there to pick cherries, apples, and pears. He would wade across the stream that once flowed throughout the orchard but now was obviously dry, with everything around appearing dead. He explained that the farmer there had mentioned to him one day that he was going to divert the stream away from the orchard to some grazing lands on the other side of the hill where his cattle could feed on the green grass and drink the water.

My father asked if that would harm those great trees, not having the stream of water. The farmer laughed and said the trees no longer needed the stream. Because of their colossal size, he said, the roots were deep into the soil, and they could get water from underground streams. After the leaves had fallen that fall, the farmer, along with his two sons, dug a new trench for the stream, which allowed the stream to be diverted toward two grand meadows that soon became green and supported many cattle. The orchard remained bare. The leaves never budded again. You see, there were no underground streams in that area. Without the life-giving waters, the great giants died and slowly started to crumble and decay.

So, it is with a woman's influence and leadership. No matter how strong her husband, sons, and daughters become, they need her sweet waters to stay alive and strong. A good woman is like the life-giving waters. Her presence, purity, beauty, and willingness to give of herself allow those around her to flourish. When a woman comes of age, she stops thinking as a child, and her life takes on a new perspective. When she matures, as if coming of age the second time, from thirty-two to forty-six years of age, she now hopefully possesses the wisdom that comes with experiencing life. She has also attained a position where a woman's rewards are beginning to be realized. She must be cautious and try to discover how to remain in this prime life stage so that she may stay wise, healthy, and alive.

A woman's total body provides the first impression to all who see her. Correct weight, cleanliness, exercise, and proper nutrition are equally as crucial as her spiritual life, in which she can communicate with others, God, and nature. She should be able to maintain a prime living for three or four more decades. The maintenance of proper hormone levels is critically important. A woman can meet up with many adversities in life and still be successful if she has a solid foundation of physical, mental, spiritual, and hormonal tools.

These disciplines must be maintained at all times if she doesn't want prime life to escape her. It takes time and effort, but the rewards of extending the prime life are great! One is impressed when one views a woman who is obviously mature, confident, graceful, wise, and with a good sense of humor. This woman has been fortunate indeed, for she has incorporated into her life mental, spiritual, and hormonal values.

Recently, a young lady came to my office complaining of menopausal symptoms, such as hot flashes, sweats, irritability, depression, and insomnia. These are not unusual symptoms for menopause, except this young lady was only twenty-four years of age. Taking a medical history revealed that her mother was very low in thyroid and had menopausal signs and symptoms by the time she was thirty-two years of age. Also, it was noteworthy that this young lady was found to be hypothyroid, that is, having low thyroid hormone levels. She had severe premenstrual tension, a hormonal deficiency of progesterone, and often skipped periods. Her hair was dry, and she had acne.

Upon physical examination, she was found to have underdeveloped breasts and infantile-type genitals with very little pubic hair. Even though she had been precocious, she had had very little enjoyment from lovemaking. Pap smear, maturity cell index, and a blood test, all confirmed low estrogen or a low female hormone level. She informed me that she had a stormy time as a teenager and quit college, even though she had been an excellent student. She ran away from home and lived in another state for three years. She had emotionally and spiritually hit bottom. Drugs had finally taken her so low that she was afraid she would overdose and die. She called her father and asked for his forgiveness and another chance in life.

The patient's father had heard several of my lectures, and he told her she could come back on one condition: to make a doctor's appointment in our clinic and then agree to follow the therapy prescribed. I don't usually babysit wayward daughters of wealthy industrialists, and when I do, I advise psychiatric or psychological counseling in addition to endocrine therapy. A woman who has been low in hormones for a very long time will usually be affected by a very negative self-image.

They usually seem to be in a rut. Hormonal deficiency results in erratic and irritable mind states that make otherwise happy, productive thinking or conclusions impossible. We prescribed oral and injectable estrogen and progesterone for this young lady.

By her third cycle, she became regular in her menstrual flow and periods. She no longer had premenstrual tension. Her hair and skin looked beautiful again, and more importantly, she soon became a beautiful person inside and out. She was an exceptionally gifted and spirited woman who could survive life's adversities intact. Her Only problem was having her hormonal deficiency corrected. She soon blossomed like a flower. To know her was to love her. Her friendship was a blessing to all who knew her. Unfortunately, this example shows the exception rather than the rule.

Usually, the man is the breadwinner in a family. He may even set the image for the family, but invariably, the woman sets the entire aura of the family, just as the sex hormones set her whole mood! She is usually the influential central figure. Therefore, when you extend the prime life of a woman, you are affecting the lives of many people, including the husband, children, neighbors, friends, relatives, and business contacts. Formal business contacts are usually business people directly or indirectly related to a woman's career or her husband's career, or a joint venture. Informal business contacts include any activity outside the home.

"My inspiration, my companion, my best friend" are some of the comments that husbands use when referring to their wives. The ability of a wife to be compatible very often depends upon adequate endocrine therapy.

Ask any man or child what happens to their family if the mother had a hysterectomy or went through premature menopause and received no help for endocrine therapy.

I'm not saying that the woman who has always had a dagger for a tongue will be transformed into a gracious belle of the Great

South, as she was known before the Civil War. I am saying, however, that even if other factors are all on "Go" for a woman, the decline with an inherently low level of hormones can cause her to sink quickly. She soon becomes a hollow vessel. She will rationalize that her goals were unobtainable anyway. Every woman has the ability to set goals and to work towards those goals.

Most men do not inherently dream, plan, or react with imagination without the positive influence of the woman in their lives – the mother, the wife, or even a daughter.

Chapter 2: Pioneers Of Hormonal Replacement Therapy

An early pioneer, Dr. Fuller Albright, was the first doctor who was brave enough to come out and state that estrogens and progesterone could be prescribed for many different symptoms, including depression, vascular diseases, osteoporosis (weakening of the bones), and other ailments common to women. His great outlines were followed by other scholars, including Dr. Wilson, who wrote Feminine Forever in 1967. Another student of Dr. Albright was Dr. Greenbalt, who was internationally known as an Endocrinologist, a Pathologist, and an OB/GYN Specialist. His cohort, Dr. Gambrill, has written many articles regarding the administration of estrogen and progesterone. Pioneer Dr. Kistner of Harvard, the head of OB/GYN there, has also been watching this controversy for over thirty years.

An interesting sidelight goes back in history to a chemist, Felix Maxwell. Maxwell was an eccentric American genius of science and chemistry. Being temperamental, he would often leave the job angrily, not giving notice or a reason for quitting. During one of Maxwell's trips to Mexico, according to old Indian folklore, he thought he found a substitute for the making of progesterone. At that time, progesterone was incredibly expensive. One could not afford to use progesterone clinically. A small jar of three ounces of progesterone would probably cost one-quarter of a million dollars.

Maxwell thought he had proof from the old Mexican-Indian medicine folklore that he could make progesterone out of the roots of some wildflowers, similar to yarns in America. The Indian women had used this ingredient for a good many medical reasons, primarily for irregular bleeding, menopausal hot dashes, depression, and pains in their bones. Maxwell lived among the Indians of Mexico for several months, living and working out of a makeshift hut. He encouraged the Indians to gather yarns and roots and processed them in a very crude manner in his hut. After Maxwell discovered that he had, in fact, produced progesterone, he went immediately to Mexico City, where he made contact with a new chemical company that two Hungarian immigrants had started. They both instantly realized the significance of producing progesterone. A jar of less than one pint, measured out for consistency, was valued at over three-hundred fifty thousand dollars.

An agreement was made, patent rights were placed on it, and production was started. Later, the company was purchased by a pharmaceutical company in America. They provided Maxwell with a new research laboratory, giving him many fringe benefits and allowing him to continue his brilliant research.

However, three months later, Maxwell sold his stock in the pharmaceutical company. He walked out of the laboratory, never to return. Large pharmaceutical companies all over the world have since been looking for him. No knowledge of his whereabouts or whether he's even alive is known to date.

The history of hormones may similarly parallel this tragic but comical story of Maxwell – the eccentric scientist. Finally, researchers may have found a proverbial fountain of youth. However, because it caused fear, it became berated and falsely attributed for causing cancer.

In actuality, progesterone is the preventer of the dreaded disease of cancer, even when administered incorrectly. The fear generated in the late sixties and seventies is still present in the eighties. Doctors, for one reason or another, consistently tell patients that hormones will cause cancer. It must be assumed that many doctors have not read recent literature to learn otherwise.

Recent reports have been published regarding the twenty-year studies that have vindicated estrogen and progesterone from being considered cancer-causing agents. In fact, these research studies have indicated a decrease in the incidence of cancer of the breast, ovaries, and uterus by over fifty percent in women taking estrogen and progesterone. A survey in Southern California regarding how to administer correct hormonal treatment found that less than eight percent of medical doctors are current in their knowledge of hormone therapy. Due to fear of the litigious nature and the breakdown of patient-doctor relationships, less than half of the eight percent of the doctors who are knowledgeable will even admit to knowing how to prescribe hormones. In fact, those very doctors usually deny patients any hormone therapy whatsoever. The philosophy at this time is one of extreme caution and avoidance of this subject.

Even though it has been shown that hormone treatment will prevent half or more cancer cases, some women will contract cancer regardless of hormonal therapy. Therefore, until the final cure is found, most doctors would rather be silent than sued. However, women have indeed sued doctors for "causing their cancer" when, in fact, the doctor administered the hormones correctly, and the patient would most probably have succumbed to cancer anyway.

Dr. Eugene Corley is a highly respected and honored physician in Connecticut, where he has practiced for over forty years. In his early days of practice, when I was a young lad, he started his practice in our home. Our living room was the waiting room. My earliest memories were of my father's intense interest, compassion, and study of the illnesses of his patients.

In the early 1930s, my father decided to complete his residency in Psychiatry while waiting for another position to open for continued residency in the General Hospital in Boston, Massachusetts. In those days, a residency could include Neuro Surgery, Orthopedic Surgery, and General Surgery in two years. During the first year, he decided to take a residency in Psychiatry due to a higher pay. However, it was an unpopular residency because few medications were known to assist psychotic people.

Upon entering the Connecticut Psychiatric Institute (mental institution) as it was called in those days, Dr. Corley Sr. noticed that there were eight times more women than men in the wards. Most of the women had been admitted at about the time of menopause, and many had remained for the rest of their lives. He asked the long-time resident pathologist about a large number of older women. The pathologist replied that he was glad to finally find someone with such keen observation skills who was also perceptive. He continued by saying that women have a difficult time in life without conveniences in the kitchen or the safety of childbirth. A woman could be physically worn out by the age of thirty-five or forty-five, at which time Mother Nature would pull the rug out from beneath her by diminishing her hormone level.

On top of all of the other physical problems, many of the women became psychotic at that point. Depression, which accompanied this fall in hormones, was persistent, recurrent, and severe. Emotional problems led to family problems. Poor nutrition added to the mental upheaval, and physical disorders followed in their course. The two doctors agreed that a need for hormone replacement could cause these symptoms. They decided to request permission from the State to allow the administration of hormones to these women with the possibility that they could be discharged. However, realizing that it was during the middle of the depression, they knew the State would not allocate money for experimentation purposes. My father was troubled by this fact and could not sleep that night.

He walked the wards in the insane asylum until a thought suddenly struck him. At 1:30 a.m., he called the hospital administrator and said he had an emergency. "You must come immediately," he demanded. The administrator protested. Later on, the administrator admitted that he thought that Dr. Corley, himself, would have been better being a patient in the hospital rather than a physician. Nevertheless, the administrator drove to the hospital, where the two of them poured over books and research studies until about 6:00 a.m. that morning. The two doctors called their mutual colleague, a pathologist, and requested that he meet them for breakfast at 6:30. Dr. Corley indicated to the pathologist that the hospitalization expenses incurred to house these women for another two years would fund even the most expensive treatment of hormones. The doctors believed that within two years, they would be able to discharge at least ten percent of the women in the ward. The State of Connecticut would be financially rewarded.

Excitedly, they sent a letter to the State officials. By return mail, the State officials allowed a one-year program rather than the two years requested, in which to discharge ten percent of the patients. Otherwise, the experiment would need to be discarded.

The experiment was started on schedule, and by the end of the first year, more than ten percent of the women were markedly improved and released. Dr. Corley then moved on to his residency in Boston, Massachusetts General Hospital. A letter from the partner doctor came to his attention, stating that they could discharge over thirty percent of those women back to their homes after eighteen months. Bizarre homecomings took place as women who had been incoherent, amnesiac, and psychotic for twenty years or longer were returned to their homes with normal health again. It was almost as if some of these women had been reincarnated. Many of their families and friends had given up hope.It was a miraculous turn of events in the lives of most of these women.

Little wonder that I decided early in my life to follow in the footsteps of my father and other pioneering physicians. As a practicing physician myself, my study, research, practice, and questioning led me to realize that most women could be treated for the fundamental cause of their ill health. Following the study and research of reputable pioneers in the field of hormonal therapy and observing the effects of hormonal treatment on my own patients, I have reached conclusions, which I have further outlined in this book.

Chapter 3: The Epidemic Of Fear – The Trial And Error System

The 1930s and 1940s marked the dawning of vast scientific discoveries in the fields of science and industry. Medical science particularly saw the beginning and the birth of man's right to understand and use hormones.

Through some lucky discoveries, including Maxwell's discovery of progesterone, by the early 40s, it soon became within the financial reach of patients to receive hormones as prescribed for various illnesses. It was soon discovered that the use of hormones could be compared to the fountain of youth. It was also found that hormones were the world's most natural tranquilizers. The use of hormones seemed to have a host of beneficial effects for the body that couldn't be distinguished in those days except to say that they certainly kept a woman younger, healthier, and happier. Their use continued, and it soon gained in popularity. The main thrust of usage was in prescribing estrogen tablets, which at that time were called Premarin. As the popularity of Premarin and some of the other orally ingested estrogens grew, there began to be some side effects and adverse reactions. An occasional cancer would be detected, and threads of suspicion started.

Maybe, it wasn't a good idea to use estrogen by itself, but nevertheless, the fad continued. In fact, it became so popular that the next man who entered the scene who really caused an upsurge medically was a man by the name of Dr. Robert A. Wilson, M.D., a well-known OB/GYN specialist and a student of Fuller Albright, M.D., at Harvard, who was a proponent of the use of hormones.

Dr. Robert A. Wilson's Contribution To The Welfare Of American Women

Dr. Robert A. Wilson was a student of Feller Albright, M.D., Professor of OB/GYN at Harvard. What an outstanding professor and teacher Dr. Albright must have been! He inspired other great men, including Dr. Robert Greenblatt, who made their mark on medicine and continued to be the steadfast pillars that we cling to regarding the use and abuse of hormones. Dr. Wilson became well-known for his work and belief in hormones. In fact, he wrote a book that became a bestseller in 1967 — "Forever Feminine," in which he laid out the tragedies that had occurred needlessly in and around his family, as well as in and around the families of his patients.

Dr. Wilson stated that he saw useless tragedy and sadness taking place, which could have been averted if these women had received the proper hormones. He went on to describe the appropriate use of hormones. He outlined how women should be cyclically placed on estrogen and the other hormone progesterone, also prescribed according to a woman's cycle. Dr. Wilson was years ahead of his time. He was already calling the shots correctly, and his methods have turned out to be classic works of art. It is now known that the safe and best use of hormones can be brought about by cycling estrogens with progesterones.

"Forever Feminine" is not to be confused with the sexuality and sensuality of Kathleen Windsor's "Forever Amber." The essence of Dr. Wilson's book is more in tune with the idea that, given proper hormones, women can stay young and sexually attractive. When "Forever Feminine" was published in 1967, it was will-read. In fact, it was a bestseller. It seemed that every woman in America who read the book took the message literally as if she had found the Fountain of Youth. Many women presented the book to their doctors and pressured them into prescribing them hormones.

Unfortunately, more women read "Forever Feminine" than physicians. Their doctors would then place the women on estrogen alone, as he was used to doing,which brought about tragedy. The tragedy resulted in Dr. Wilson being harassed by the County Medical Association and having the threat of losing his medical license. The emotional trauma was too much for Dr. Wilson, and he retreated.

The story of Dr. Wilson reminds me of Dr. Semelweiss. Dr. Semelweiss was a quiet Australian doctor who spoke to other doctors in a kind manner and said, "It seems to me that if we would wash our hands between pelvic examinations of pregnant women, that less of them would die of infection." Years later, the bacterial infection theory was upheld and proved to be true. Unfortunately, Dr. Semelweiss did not live long enough to realize his discovery. His peers were envious of him; some hated him. His medical license was taken away, and eventually, he left his country. Dr. Semelweiss died in another country, broken financially, with his spirit did not know that he had made one of the greatest medical discoveries of his time.

The Cancer Scare Of The 70s

An old saying, "figures can't lie, and liars can't figure," is not entirely true. The statistical error was unknown at the time, but cancer of the uterus and the use of hormones were held in contempt for many years by doctors who were in mainstream medical practice.

It was not known that cancer of the uterus was a slow-growing cancer. In fact, this type of cancer could grow slowly and insidiously in the womb from two to six years before bleeding. In fact, bleeding was a very late sign. Often it was too late for a woman in her sixties or seventies who would suddenly have a period. Cancer had already spread beyond the uterus walls by the time any bleeding appeared. During the first few months or years of having cancer present in the uterus, if a woman receives estrogen, it will cause immediate bleeding, which is then considered to be an advanced signal of cancer. Left up to the cancer to bleed by itself, it bleeds too late for treatment. Some women who already had cancer of the uterus were inadvertently given hormone tablets or injections, or both, and they bled almost immediately. Left alone, this takes two to six years to trigger these statistical occurrences. It appears that the cause and effect would indicate that estrogen caused the cancer. This statistical error stands still today.

A woman who is subjected to unopposed uncycled estrogen will experience a change from the normal endometrial lining of the uterus to a hyperplasia or overgrowth of ultimately abnormal cancerous cells. This type of cancer only occurs in a woman with endogenous estrogen, where the estrogen within her own body is too high. Being too heavy or post-menopausal sometimes causes fat cells to produce a hormone called estrone.

Estrone is a metabolite produced from the fat cells after it has utilized a more or less neutral adrenal gland hormone called Androstenediol. A woman who is fifty pounds overweight risks nine times the chance of getting cancer of the uterus than a woman of normal weight. Her fat cells erroneously produce a high dose of estrone. A doctor prescribing estrogen uncycled to a woman of any weight can cause the lining of the uterus to become erratic and develop hyperplasia with a heavy thickening of the lining of the uterus. This condition is not cancerous by itself, but if continued, it can become like adherent cells. Finally, cancerous cells will form in women destined or prone to getting cancer due to genetic factors. Another factor, believe it or not, besides being genetic, is a mental attitude. Statistical errors are generally found whenever the cancer bled at the same time when estrogen treatment was introduced.

The Wrong Statistical Conclusions

The wrong statistic was drawn when greater attention was paid to women's gynecological cancer-related problems. During the late sixties, more frequent cases of cancer were found, and that paralleled the usage of the hormone estrogen. Statisticians noted increased usage of hormones and the increased incidence of cancer simultaneously. They drew erroneous conclusions that hormones are the primary cause of cancer of the uterus. They couldn't have been further from the truth. As a matter of fact, as will be indicated later, the use of estrogen and progesterone, both female hormones, when given cyclically, actually protect a woman against cancer of the uterus.

Malpractices

The definition of malpractice is when a doctor practicing is more fearful of malpractice rather than being free to practice single-mindedly for the patient's benefit. Many studies were conducted in San Francisco, where Melvin Belli, a practicing attorney, is probably the world's greatest and yet least-known malpractice attorney. Four out of the first seven studies in San Francisco concluded that hormones caused cancer. The pathologist often looked at a slide and could not discern whether the cells indicated severe hyperplasia caused by giving estrogen alone or whether this was an initial stage of cancer. The thought occurred that if one case of cancer was overlooked, then Melvin Belli would be in a rage.

Calling it cancer, even if by mistake, would make everybody happy because: (1) the patient will be told she has early cancer and needs a hysterectomy. She will be frightened, but after the hysterectomy and five to ten years, the doctors will indicate that she is absolutely cured. The pathologists would be happy because they wouldn't get sued, and the patient would be satisfied because she believes that she has been absolutely cured of her cancer. The surgeon is also happy because he performed a hysterectomy that saved her life. Later, the ten-year study was repeated, and statistically, there were too many women alive to have had cancer of the uterus. The original slides were reviewed. Pathologists who were not subject to malpractices soon discovered that over thirty percent of the slide smears, which had been called cancer, were found to have never been cancerous in the first place.

Hyperplasia In Cancer Cells

In my opinion, malpractice is an epidemic that is growing. Only a few men, such as Dr. Stalwart, who has plunged ahead, calling the shots as he sees them. He does not practice in the unknown world of fear. Without the pressure of fear, it is easier to discern the difference between severe hyperplasia or pseudo adenocarcinoma, which is not a true carcinoma, or early-stage carcinoma, which is true cancer. One can certainly distinguish between severe hyperplasia and cancer. Most of the pathologists currently trained, however, are often confused. The training was not presented to them regarding the differences. It has not been known that women ages fifty-five to sixty could have hyperplasia of the endometrium without being in the early stages of cancer if they have not taken it upon themselves to become educated in this area. When in doubt, the knowledgeable physician will send the slide smears to a few chosen pathologists to determine whether it is a questionable case of hyperplasia or actual cancer.

True Allergic Reaction

Occasionally when women take estrogen or the birth control pill by mouth, they show a sensitivity which results in an inflammation of the endothelial linings of the blood vessels of the leg. With enough inflammation of the blood, clots will form along the walls of the arteries, and then there are two dangers. The first danger is the impaired circulation of the leg. The other severe reaction is when small parts of these spots break off and travel from the leg veins into the bloodstream. If these small spots or clots reach the heart or lungs, they may cause a severe disability or even death. Only a few women have had severe allergic reactions when higher doses of birth control hormones are prescribed. These cases are very rare. There is some belief that these reactions are related to smoking, although it is too early to make that determination. Most women do not have a tendency toward thrombophlebitis unless they are severely obese or have been smoking.

Out of deference to women who are truly allergic to birth control pills, everyone placed on birth control pills should watch for any redness, inflammation, soreness, or muscle spasms in the legs. She should discontinue using the pill if any symptoms occur and see her physician immediately. Unfortunately, women who developed problems ignored the redness, inflammation, soreness and/or swelling of the legs as severe symptoms.

Cardiovascular Arrest And The Pill

There is currently agreement that women who are likely to have a stroke due to taking birth control pills are ones over thirty-five years of age and who smoke.

Women who are over thirty-five and smoke should be warned against taking the birth control pill. They should be given a choice, give up smoking or the pill. Smoking by either sex has been proven to cause a greater chance of cardiovascular problems. I personally believe that smoking causes more adverse effects than age. After the age of thirty-five to thirty-six, hormones start to decline in most females. In my opinion, women over the age of thirty-six should be placed on another type of estrogen therapy with larger doses, whether tablets or injections. This would provide further protection against any cardiovascular problems. Women over thirty-five frequently had difficulties for two reasons. The birth control pill gave too small dosages to protect against cardiovascular problems, and smoking over many years caught up with them.

Erroneous conclusions can be drawn from statistics. I remember reading an article written to physicians indicating that one statistical investigator, in his article titled "The Pill As A Cause of Pulmonary Cancer," proved that the dangers of the pill were far-reaching. The statistician studied many hieroglyphics and found that with the greater use of the pill, there was an equal rise in lung cancer. The conclusion pointed to the pill as the causal factor for increased lung cancer in women. The women's liberation movement allowed women to smoke more freely in public. More women, in fact, do smoke than was previously realized. Public health statistics now indicate that the increase in incidents of lung cancer in women is due to their smoking habits. Smoking, as a causal factor of cancer in women, is becoming one of the greatest killers of women.

Smoking excessively over a period of years while taking the birth control pill is now being shown to be the genuine causal factor in the increased statistical rise in lung cancer in women.

Blunders And Consumer Group Pressures

Manufacturers of estrogens have been required to place inserts or pamphlets into each package of pills. The information stated taking estrogen was a very good treatment for the prevention of osteoporosis, related complications, and even death. Dr. Jekyll and Mr. Hyde condition was created in the very next statement on the insert information. It states that your chances of contracting cancer of the uterus will be increased five to fifteen times. This erroneous blunder enraged OB/GYN Specialists in the OB/GYN Association, who sent specialists to testify before the Federal Drug Administration sub-committee hearing. Ralph Nadar's Consumer Group advocates using these statistical errors for conclusions were exposed and accused of being more interested in their political turf than the welfare of the people. The OB/GYN Association of America has pursued several frustrating attempts to discuss this problem logically with the FDA. Presently, there is a lawsuit pending against the FDA that will require an injunction against them for placing damaging and misleading information into each package of pills.

The FDA Blunder In The Case Of Medroxyprogesterone

Medroxyprogesterone has been used for years, with thousands of women taking it for birth control. The latest statistics show that medroxyprogesterone is not responsible for causing cancer but that it contains a protective factor against cancer of the uterus. Consumer groups dug up studies where massive overdoses of medroxyprogesterone were given to monkeys. The monkeys showed a high incidence of cancer of the uterus. Injections to women for the contraceptive immediately became suspect, even though, in fact, medroxyprogesterone has been used by leading researchers to reverse cancer of the uterus.

Ama – The Politics And The Science

There is a statement for both worlds. The American Medical Association often does not represent the majority of the physicians in the United States. In fact, frequently, it goes counter to the majority of physicians in the United States on many issues. It has become like any other bureaucracy with pressure and politics, lobbying on its own and becoming a political entity. There is no alliance, it seems, either to the good medical care of patients or the physicians who give the medical care. In fact, it often interferes with those efforts. A dwindling percentage of physicians continue to be involved with AMA. An increasing percentage of doctors do not want to be affiliated with the AMA. The AMA is in a position where they feel they have to make a statement that will, in some way, appease the physicians and be politically sound, even though a medical error is committed. The AMA issue of January 21, 1982, stated that, in essence, according to summaries of the 1979 report, the benefits of estrogen therapy were found to be elective in relieving osteoporosis, vaginal atrophy, hot flashes, arteriosclerotic heart disease, and possibly even psychological problems. However, it states that taking estrogen may cause an increase in cancer.

They state that estrogen significantly increases the incidence of endometrial cancer to appease the physician. However, this type of endometrial cancer caused by estrogen does not cause an increase in the death rate. So far as we know, there is no other cancer in the world that doesn't cause an increase in the death rate except for this one, founded by the AMA. It seems to be from a crazy world of politics.

The American Cancer Association has recently proclaimed that women may go three to five years without a regular Pap Smear. Doctors who investigated this, including Dr. Reichhardt, a well-known pathologist, said this statement is true idiocy. This was already tried in Great Britain, where they instituted the edict that women should only get Pap Smears every three or five years, and the incidence of cancer of the cervix increased by one hundred percent.

Dr. Vernon, former head of the Department of Endocrinology at Baylor Medical School in Texas, started saying that rheumatoid arthritis in the female and possibly even in the male had more causal factors. Probably, he said it was even being misnamed by first calling it rheumatoid arthritis. It should have been more correctly called a hormonal deficiency that causes osteoporosis.

The characteristics are specific and should be called rheumatoid in nature. Now to prove this point, he gave sex hormones to individuals having rheumatoid arthritis. He discovered that around ninety-two percent of people with rheumatoid arthritis, many having had it for years, responded in a very positive manner to the injections of their respective sex hormones. The majority of these responded beautifully. Their pain and even some deformity-type symptoms were completely relieved. Dr. Vernon's reward for this great discovery was similar to how Dr. Wilson and Dr. Semelweiss were rewarded.

The American Arthritic Association brought proceedings against him, as did his own peer group, the County and State Medical Association, and other health bureaucracies. The man's license to continue practicing medicine was at stake, almost to the point where he had to practice underground. He spent all of his time defending himself in lawsuits and had little time left to practice medicine. Instead of honoring Dr. Vernon as a hero for finding a treatment for rheumatoid arthritis, he was treated as if he were a Charlatan or a quack. However, it seems that the shoe may be on the other foot. Very recently, statistics of twenty-year studies were released regarding results on people who had been on hormones for twenty years. Lo and behold, there was a reduction of about ninety-two percent of the statistical incidence of people who would have suffered rheumatoid arthritis.

They did not get rheumatoid arthritis when they were on sex hormones. At this point, we wish Dr. Vernon good luck and hope his word will spread far and wide to physicians so they can treat rheumatoid arthritis effectively.

"Malpractice" – From Fear, Anger, And Suspicion

The late sixties saw the beginning of a climate of fear, anger, and suspicion just before the actual cancer scare. I recall the professionalism with which physicians used to be looked upon. They also worked to preserve the rights and the honors which they had. My father's dedication to medicine was admirable. Some physicians looked at my father contemptuously as if he were a workaholic. They played golf, traveled, and attended extensive alcoholic "grand slam" parties. It was difficult for me to understand whether my father was a naive country boy who became a physician who was totally in love with medicine or was truly an example of a 'high priest' in medicine who wholly dedicated himself to medicine. He wanted to help people. I learned the philosophy of medicine from my father. My mother also had a significant influence on my outlook and philosophy. She was one of the few Northern Baptist female ministerial seminary graduates. She and my father were very unusual and unique in their dedication to mankind. Together, they worked to help patients. She assisted with counseling and my father with medicine. They taught me the true meaning of psychosomatic medicine before I ever heard the word or learned how to spell it correctly.

We lived in a three-story house on Elk Street downtown in the little village of Franklin, with a population of about thirteen thousand. Everybody knew everyone. With six children in our family and twenty-three more exchange students adopted by our family, we eventually had to move up to Miller Park, the estate area, because we needed dormitory sleeping areas. All the rooms on the first story were utilized as my father's medical offices, waiting room, and treatment room. The second and third story was about where we lived. The first back room on the first floor was a very large room, a combination of a dining room and kitchen. That room was the hub of family life. In the evening, we moved upstairs to the living room where we would read, exchange conversation, listen to Mother and Father talk, and in general, grow in love and instruction from them. I remember one day when my father called my mother and said, "Alice, would you come for a minute? There is a patient with whom I want you to help me."

I would hear my father say to his patient in his big booming voice, "Now, John, you know my wife, I want you to talk to her about this problem because there is no medicine that I can give you that is going to settle your high blood pressure as long as you get upset every time you see your son. You and I both know that your son has an alcohol problem. This is very hard for Pennsylvania Dutchmen like you and myself to accept. I think you will have to learn to love him despite how he is. Be accepting of that fact. Every time he comes around, and you blow off steam, you blow your blood pressure up, and that can be a very deadly habit. I want you to stop taking those blood pressure pills which are giving you the bad side reactions you came in to see me about. I want you to make an appointment to see my wife two or three times each week for the next few weeks. See me in three weeks, and we will see how you are doing."

My father would usher John into one of those big easy chairs that surrounded the pot-bellied stove in our back waiting room just off from the kitchen. John would settle into that big chair. He was a big-boned, rough-looking, kind-hearted farmer. He was a little apprehensive about how he felt about talking about his personal problems to a lady. My mother was, and is, a short, dynamic, vibrant, and beautiful person. She has black off-brown colored hair, which is now turning gray. She was and still is a very beautiful person. She has a wonderful love for people, a tender voice, and a fantastic way of helping people. She is Portuguese and Greek.

My father, on the other hand, is Portuguese African American – with brown eyes, black hair, and skin that tans very easily to a very dark color. It is pretty humorous that when anyone comes into Upstate New York and hears one of my parents is Portuguese, they always walk up to my father and start a full-fledged conversation in Portuguese, of which he does not understand a word. At any rate, Big John would settle into the chair by the pot-bellied stove, a big leather, overstuffed chair, and my mother would take off her apron in the kitchen, push her hair back behind her ears, and up on her head, and then she would walk in and say, "Well, hello there, Mr. Brenner, how are you today?" Of course, he would appear very nervous and say, "Why, just fine, Reverend Corley, I am doing fine." And then she would take a chair, pull it up beside his chair, sit down, look at him and say, "You know, all my husband really wants me to do is talk to you about how you can control your blood pressure by controlling your thoughts and making them more restful to your nature. In other words, you see peace in your mind's eye." This woman already had a little inkling of how healing visualization and imagery worked. "John," she continued, "it will come to pass when you imagine it. I am not saying that as a minister speaking: however, there are passages in the Bible that tell us, "As you shall think, you shall become." I believe it to be true because I've seen this happen with many, many patients. It is no difficult task at all." She continued, "Do you pray very often, John?" "Well," he would say, very uncomfortably, "sometimes at slipper, but usually at Christmas and times like that." She gently urged, "Well, maybe you pray more than you realize. There is really no formal format that has to be followed in praying, one can pray with their own private thoughts without saying a word, or one can say words audibly so they can hear those words. It has a sort of double impact on our senses.

But at any rate, the most important two times of the day that you can make an impression upon your subconscious memory patterns or your habit pattern of thinking are at the first few moments when you wake up in the morning and the last few moments before you go to sleep. I want you to do this whole exercise, John. I want you to spend about five or ten minutes when you first wake up in the morning just being grateful that you are waking up without severe pain or severe disability. Be grateful for all of the blessings you have. Think of something peaceful in your life, such as the meadow scene at the farm, how you enjoyed fishing when you were a young man, or how the sun looks when it is settling over your farm. Then go ahead with your daily activities. When it is time to go to sleep, continue to picture in your mind those peaceful images and go to sleep thinking about pleasant thoughts. Be grateful for possessing those memories. Memories are true treasures that we keep with us. After you have practiced that for about four or five days, do the same procedure in the middle of the day. When it is quiet, and you are alone, spend at least ten minutes thinking similar positive thoughts by yourself. Close your eyes and fix your gaze on something very peaceful, and think about the fact that you are happy to be in such good health and live without pain and disability. Think of a pleasant scene that is your favorite spot. Also, think about sharing that favorite place or experience with your son or daughter.

Picture your son as he was when he was a little boy and before he had these problems. Accept that real part of him and love that part of him, despite feeling uncomfortable about the present issues with him. Still loving him, return to visualizing the place by yourself again. Review how grateful you are for the blessings that you have. Open your eyes, stop gazing at that area, and go about your business."

As I said before, my father would turn these non-responsive patients over to my mother, she would give them counseling, and they would return in three weeks. My father would take their blood pressure and comment, "Well, John, you did a good job. Your blood pressure is normal, and you did it yourself. You see, you really don't need me at all. YOU needed to find that part of yourself where the greatest physician of all can bring about healing."

Many times, physicians in those days had mystic insight into the fact that there is a little bit of God or a great physician within each of us. Physicians then tempered their treatment in the art of medicine to a large extent with their ability to help man see himself as he really could be. More and more, I have noticed that procedures have become a very matter of fact as we have become technical in our approach. Personal needs become emphasized less. The problem of multispecialty fragmentation started, and finally, arrogance, rudeness, and aloofness crept into the system. Patients were intimidated, embarrassed, and often angered by these tactics. It gave a black eye to the medical profession.

Soon after, the great malpractice crisis followed. Then the cancer scare, drugs, problems, and all other circumstances compounded more problems. Physicians were accused of not upholding the trust or the dedication which they had enjoyed in previous days. They were angered and hurt by this change in the attitude toward this new, rich, young breed of physicians. They had really become prevalent; there were periods of financial reversal, as is going on right now. Presently a physician, especially in California, giving hormones correctly can be sued by a patient who has cancer.

Even though the doctor prescribed hormones correctly and the hormones will protect the woman so that she will get cancer at only one-eighth the rate of women who don't take hormones at all, the doctor may still be sued.

Occasionally a woman will get cancer even though she received hormones. The fact of the matter is that her cancer will be discovered earlier, and usually, her life will be spared due to earlier detection. Now, the physician is becoming the whipping boy of the politician. The personality of the physician is considered to be aloof and appears callous on the witness stand, whereas the patient may appeal to all the sympathy of the jury members.

Jurors are more apt to relate to the patient's income and status level. Physicians who have given hormones correctly have lost lawsuits in court. Therefore, a new wave of anger, fear, and suspicion has been passed on to the doctor. The doctor will say to himself, "I don't believe in hormones because if the woman gets cancer, she will sue me, and she may win. She can't sue me for not doing anything, so I will simply refuse to give her hormones and will ignore the suffering, dowager hump, fractured hips, increased chance of stroke or heart attack, and increased chance of cancer. She certainly can't sue me for doing nothing." Unfortunately, this is state-of-the-art today. A recent survey found that less than eight percent of doctors apparently are familiar with hormonal protection. Even worse is the fact that less than half of those eight percent of physicians are knowledgeable about cancer protection with hormone therapy. Less than half of those physicians will prescribe hormones. Due to the malpractice scare, most physicians have decided to take a stand and do nothing. Therefore, the chances of being sued will be less. It is very tragic for both the physician and the patient to remain in this state.

The tragedies will be enumerated as follows.

The Tragic Statistics Of Fear (Loss Of Protection)

Because of the increased furor brought about by the press, the cancer scare, statistics, surveys, and studies done in the late sixties and early seventies, the use of hormones, both for post-menopausal women and for birth control for younger women, has been cut in half since 1975. In other words, in 1975, there were almost twice as many women getting protection from hormones than at present. Some of the statistics that have been uncovered show that for every one-hundred thousand women in the United States not using both birth control pills, there has been increased irregular periods and heavy bleeding due to not using birth control pills.

Three-hundred and twenty out of one-hundred thousand women develop iron deficiency anemia; thirty more suffer from rheumatoid arthritis, and four-hundred and forty-five contract minor pelvic inflammatory disease. In addition, one hundred and fifty-six more women are hospitalized for severe pelvic inflammation, two-hundred and thirty-five more with ovarian cysts, and one hundred and seventeen more ectopic or tubal pregnancies, often ending in death. Ectopic pregnancy is caused by scarring of the fallopian tube due to pelvic inflammation from infections. Since these hormones protect against cancer, there are five more cases of endometrial cancer and four more ovarian cancers per one hundred thousand women. Daily I hear stories of women who were on Premarin estrogen for years and were spared osteoporosis and cardiovascular diseases. But because they can't find a physician who will continue to allow them to take hormones, they now suffer coronary heart attacks, paralyzing strokes, severe pain in the spine, dowager's hump, and hip fractures. We now wonder what we can do to reverse this epidemic of fear which ends in such tragic statistics. In my opinion, it will take action from more than just doctors to change this. I think it will take the action of the patient and interested physicians to rectify this problem.

The Need For A Consumer Group

I think it has become apparent by now that people in the United States have constitutionally had the right of free choice, health, and happiness. This freedom has seemingly been curtailed recently by bureaucratic institutions. It seems that a consumer group should be formed that will protect the right of free citizens to make choices about their health and welfare. Physicians interested in providing alternate therapy services to these patients should be allowed the free choice of therapy they consider proper. These therapists should be allowed to help people learn the facts and make their choice. Those who are interested should decide whether a physician who wants research information or a private citizen is interested in hormonal therapy.

Chapter 4: Now The Good News

Medicine's Best-Kept Secret

After a twenty-year survey on the benefits and risks of the contraceptive known as "the pill," it was proven that it is time to reconsider the alleged risks and provide this information to the public. Citizens will discover that this method is safer than they thought and is suitable for a larger percentage of female patients than previously believed. In reviewing the benefits of the birth control pill, it is safe to say that it is the most elective and convenient method of contraceptive currently available.

Three major myths caused the scare for the public as well as the physicians over the last ten to fifteen years. The first myth was that there would be an increased risk of developing cancer. Many studies over the past twenty to thirty-five years have proven a new conclusion. Depending upon the population and how long each group kept on hormones, the risk of breast, uterus, and ovary cancer in women taking the correct combination cyclically of estrogen and progesterone was reduced to between one-half and one-eighth of those women not taking hormones. Stated another way, women not taking hormones face cancer of the breast, uterus, and ovaries two to eight times more often than women taking hormones cyclically in the form of birth control pills or cyclic estrogen/progesterone treatments.

The second myth was that women feared that the birth control pill, which made them appear to be "sterile" for a very long period, would cause them to remain sterile permanently. The opposite, in fact, turned out to be true. Earlier, I stated that women who took the birth control pill were protected against irregular periods and infections of the pelvis. Abscess, including inflammation of the tubes and severe PDI, often required surgery and frequently caused the women to be sterile. The benefits of oral contraceptives seem obvious. Protection against infection can be explained more thoroughly. When estrogen and progesterone are given to women in the form of a birth control pill, it causes a mucous plug to form at the neck of the cervix, which is the neck of the uterus.

This thick plug results in little or no bacteria being able to get into the vagina and into the uterus during intercourse or on the tubes, thus infecting the tubes and even the entire pelvic area. Women who have never taken birth control pills are prone to ectopic pregnancies.

Ectopic pregnancies provide the highest incidence of deaths in teenagers. These pregnancy-related deaths would have been avoided had the women received the protection of the cervical plugs, which are formed when taking hormones. Promiscuous girls end up with repeated infections which develop scars in the tube or tubes and usually cause sterility. Ectopic pregnancies are caused because the egg can move only partially within the tube. A tubal pregnancy has its inherent danger of exploding as it develops for several weeks, thus causing a ruptured tube and blood vessels. Some of these girls bleed to death before they ever reach the hospital. This problem should dispel the fears of remaining sterile after taking birth control pills. No such danger exists to the degree that is worth mentioning. The dangers, however, of not taking birth control pills and remaining sterile are worse because of the scars from infection, which is prevalent and becoming more prevalent as fewer women take the birth control pill.

The third myth that frightened women away from taking birth control pills or hormones was the fear of heart attacks and strokes. These worries are quite unwarranted. The only women who really suffer heart attacks and strokes are women who are over thirty-five and have had a history of smoking cigarettes. Smoking should be strongly discouraged, and a woman who is over thirty-five and wishes to continue birth control pills could probably take them safely if she would not smoke and has no history of blood or vascular diseases. The studies that were done in the United States as well as Great Britain have shown that women taking oral contraceptives have a decreased risk of developing benign breast disease, including cystic breast disease and benign breast tumors. They are also fifty percent or more, less likely to develop cancer of the endometrium unless they were in menopause, who were taking estrogen by themselves. Again, when it concerns fertility, researchers found that the worst thing that could happen was occasionally, a few women might experience a delay of several months in the return of fertility.

Nevertheless, fertility remained good after that, and they did not have any increased incidence of abortion. There was no increase in the incidence of infants having congenital anomalies. Another discovery was that the increased risk of having cardiovascular disease was simply not present in women who do not smoke and are under thirty-five.

Recent studies on the birth control pill and hormones given to menopausal and post-menopausal women reveal there is no risk of developing strokes, fatal myocardial infractions, or heart attacks in women unless they smoke. During the last ten years, there appears to be an actual decrease in fatal heart attacks or paralyzing strokes in women who have been using hormone pills. When the studies have been summarized, it seems that women who are protected with hormones, whether in birth control pills or estrogen and progesterone, cyclically. Let's compare women who do not take any hormones for conditions that develop due to hysterectomy or menopause. It has been found that women who do not take hormones are three times more apt to develop a heart attack, five times more likely to develop paralysis from a stroke, two to eight times more likely to develop cancer of the breast, uterus, or ovaries, ten times more apt to develop rheumatoid arthritis, twenty times more apt to develop severe depression requiring therapy, and approximately two-thousand times more likely to develop osteoporotic fractures of the vertebrae. Dowager's hump and/or hip fractures. In the United States alone, hip fractures cause fifty-two thousand deaths in women each year. Another eighty-five thousand women annually are placed in convalescent hospitals for the rest of their lives. They suffer from severe pain from hip surgery, where the bones are too fragile to mend. Added to these statistics is the fact that premenstrual tension, painful menstruation, irregular periods, anemia, frequent D & C's due to irregular periods, and pelvic inflammatory infections are greatly increased when women do not take hormones.When these facts are taken into account, this is suddenly an overall picture of quantities of lives being lost due to not taking protective hormones.

The quality of life has been lost by thousands of women. Their mates and families suffer depression and pain also. Statistics climb in geometric proportions. The haunting fact is that most of these tragedies are unnecessary.

Since I've been in practice, /or twenty-two years now, there have only been two incidents of hip fractures among all the women in my clinic who have taken estrogen and progesterone. Many elderly women now live longer and happier because they are on hormones.

Now that the good news has been released about "The Pill," more women will be able to make their own free choice. They will know the true statistics and will be free from fear regarding taking birth control pills of hormones when indicated.

Those Courageous Men And Women And Their Finding

Consider the studies of Robert Kistner of Harvard, Dr. Greenblatt and Dr. Gambrell of the Medical College of Augusta, Georgia, Dr. Felix Rutledge, University of Texas, Dr. Leon Sperof of the University of Oregon, Dr. Nachtigall, Dr. Georgianna Jones of the Johns Hopkins, Dr. M. Dorothes Kerr of Cornell, Dr. Nachtigall of Belleview Hospital in New York, Dr. Georgianna Jones, professor of OB/GYN at Johns Hopkins, and many others. Their studies vary from five to thirty-five years and have repeatedly concluded in favor of cyclic hormonal therapy in the form of cyclic estrogen and progesterone.

It's time that people take into account what has been written. If hormonal therapy is controlled the right way, it will work to great advantage. Let's have our fears laid to rest by the results of research.

Dr. Greenblatt, in his ten-year study of estrogen and progesterone, showed no cancer of the uterus or breast. In another study; that he completed on one-thousand females taking hormones for over thirty years, it was found that there were no deaths from cancer of the uterus. Another study of ten years completed at the Goldwater Memorial Hospital in New York on estrogen and progesterone found no breast or uterus cancer among women taking estrogen and progesterone. Another statistic, this time from the Cancer Association, shows that in ninety-one percent of the women having cancer of the uterus, none of those ever had been given hormones at any time in their lives. So, it seems clear that women without hormones definitely have a greater risk than women taking estrogen and progesterone.

Early bleeding caused by hormones, however, led to earlier detection of cancer already present in the uterus and thus saved lives. The common denominator of cancers of the uterus that we've studied is a lack of progesterone being present, and therefore estrogen was unopposed and not cyclic in nature. Interestingly enough, other factors, as stated before, caused the increased risk of uterine cancer. It is now possible to determine which patients are most likely to be susceptible to uterine cancer. If a woman is twenty-five pounds overweight, she is designated to have three times the chance of contracting cancer of the uterus. If the woman is fifty pounds overweight or more, she will have a nine times greater chance of contracting cancer of the uterus than she would have had normally.

If a woman has had one pregnancy or less, it was probably due to a lack of hormones. This lack of hormones makes a woman more susceptible to cancer of the uterus. Suppose a patient has diabetes mellitus, liver disease, estrogen-secreting ovarian tumors, or polycystic ovaries, which often are due to low hormones. In that case, she has a greatly increased chance of suffering from cancer of the uterus. It's also essential to note statistically that cancer of the uterus has the lowest death rate of all gynecological cancers. In control groups, women receiving no hormones were found to have cancerous uterus, breasts, and ovaries, as well as a dramatic increase in death from coronary heart disease, paralysis from strokes, pain, and crippling due to osteoporosis.

Not all post-menopausal women need estrogen replacement. Some produce enough estrogens to keep them from having symptoms. Additionally, both ovaries and adrenal glands secrete androstenedione which is then converted to estrone in the fatty tissues. When ovarian estradiol secretion diminishes at menopause, the principal source of estrogens becomes adrenal estradiol, converted to estrone peripherally in the fatty tissues. The two factors which can increase and maintain endogenous estrogen levels after menopause are: (1) increased production of androstenedione and (2) the increased peripheral conversion of this hormone to estrone.

Patients with untreated adenomatous hyperplasia will find that thirty percent will develop cancer within ten years. Unopposed estrogen therapy plays a role in the development of hyperplasia and neoplasia, making for incomplete shedding of the endometrium. Giving progesterone will ensure complete sloughing and prevent the continued endometrial proliferation that may result in hyperplasia or neoplasia. In addition, progestins provide excellent palliative therapy for even metastatic endometrial cancer. Remissions lasting as long as four to five years have been reported.

Valuating Postmenopausal Bleeding

Adenocarcinoma has been reported in as many as 13.1 percent of patients with post-menopausal bleeding. Since ninety-seven percent of women treated with a combination of estrogen and progesterone have withdrawal bleeding at the end of each treatment cycle, it is not considered abnormal if it occurs on schedule. However, any unexpected bleeding should be considered abnormal.

We recommend that hormones be prescribed in the following manner: Day one through twenty-five of a woman's cycle estrogens are usually administered, and then the progestins are added from day sixteen through twenty-five. There will be a five-day rest prior to beginning a new cycle. After a hysterectomy, the estrogen will be continued alone if it is deemed necessary for a hysterectomy to occur for persistent hyperplasia.

Progestin Therapy For Women At Risk

Women with hyperplasia and predisposing factors such as obesity, low parity, liver disease, on estrogen replacement, a post-menopausal woman with a uterus (whether she is estrogen-treated or has enough endogenous estrogen to relieve symptoms), should be given a trial of progestin to see if it will induce withdrawal bleeding. The progestin hormones should be continued for ten days each month for as long as withdrawal bleeding continues. If withdrawal bleeding occurs, then a three-month trial of progestin is indicated to determine the effects of either the exogenous estrogen or endogenous estrogen. For a woman with a negative progestin challenge and no withdrawal bleeding during three months of progestin therapy, I recommend repeating the test as a single ten-day course once a year around the time of the annual examination. Patients with endometrial hyperplasia should also continue the progestins for ten days a month as long as withdrawal bleeding occurs. If curettage shows that the hyperplasia has been reversed on progestin therapy, then the job is completed. Abnormal or unscheduled bleeding must be investigated promptly by a curettage. With these methods, most endometrial cancers can be prevented in treated and untreated women. May I emphasize again that the minimum dosage effective in preventing most hyperplasias and endometrial cancers is either norethindrone acetate, five mg, or medroxyprogesterone acetate, ten mg. for ten days each month. On this dosage, only a rare patient will still develop hyperplasia.

Any abnormal post-menopausal bleeding must be investigated. Since progesterone is produced for at least ten days during the normal twenty-eight-day cycle of a normal woman, the ten-day course of progesterone is logical. Dr. Lila Nachtigall, Associate Professor of OB/GYN at New York University Medical Center and Director of Gynecological/Endocrinology at Bellevue Hospital, stated in her research report article that over four-hundred female patients at the Goldwater Memorial Hospital in New York City were screened for final selection of one-hundred sixty-eight. There was not one single case of endometrial or breast cancer among the women in her group who took estrogen and progesterone hormones. Using placebos for an untreated group, there was one endometrial cancer and four breast cancers during the ten years, which is about the national average for this age group. In addition, there was evidence of reduced cardiac risk in patients treated with hormones. The cholesterol levels decreased with hormone therapy and remained stable. However, the cholesterol increased in the control of untreated patients.

Osteoporosis appears in menopause, and it was found that patients in the untreated control group had already shown signs of this disease and now showed an increase in its severity. On the other hand, the estrogen/progesterone-treated group of women who stated their treatment after the menopause was found to have improved throughout the ten years, actually gaining bone destiny instead of losing it. After this study, one can see that a special regimen of female hormones can protect women against endometrial and breast cancer and reverse osteoporosis symptoms.

Protection Against Heart Disease

It isn't totally certain how or why estrogen protects women against heart attacks. It is known, however, that estrogen definitely lowers the cholesterol level of the blood, and this may be the major contributing factor.

Protection Against Vaginal Problems

When a woman's estrogen supply is decreased, there will be less vaginal mucus produced. As a result, the walls of their vagina will become thinner and very fragile. Generally, these changes occur gradually, while the symptoms may arise suddenly. Changes in the urinary tract and urinary infections have become more prevalent. Estrogen will nearly always remedy these physical problems.

What About Hot Flashes?

Estrogen will relieve most vasomotor symptoms of menopause, including hot flashes, sweats, insomnia, and palpitations which begin when the natural supply of estrogen wanes. Estrogen replacement therapy will relieve the vasomotor symptoms altogether for most women within a week or with possibly only two or three doses.

Emotional Symptoms

It is extremely difficult to separate emotional responses during menopause from physical symptoms. Menopause is, after all, a dramatic symbol that a woman is not getting younger. The decline in hormones produces very real effects on the emotions. Their replacement will also promote a renewed feeling of wellbeing.

Chapter 5: All You Wanted To Know About Hormones But Were Afraid To Ask Your Doctor

Will Estrogen Keep You Young?

Estrogen makes a woman feel younger. Other good health habits will assist in softening the effects of aging when estrogen diminishes naturally in the body. There is a very real effect on the skin, muscles, breasts, figure, and hair. This does not mean estrogen replacement therapy should be used solely for cosmetic reasons. There is absolutely no need to suffer hot flashes, numb fingers, or sleepless nights. Taken correctly, estrogen replacement can be perfectly safe and will go a long way toward enhancing the life of the menopausal woman.

The fact that progesterones will reverse hyperplasia to normal endometrium in most patients has been well demonstrated. For example, Gambrell found in his research that with progestogen therapy, varying degrees of hyperplasia reversed to normal endometrium in one-hundred ninety-two out of one-hundred ninety-nine patients.

Breast malignancy in the estrogen/progesterone-treated woman is significantly lower than in patients receiving no hormones. There is no significant difference in the incidence of breast cancer between estrogen and estrogen/progesterone users.

Premenstrual syndrome is just another hormonal deficiency. Currently, it is a trendy subject at clubs, debates, and women's gatherings. Professionally, within the United States, physicians seem to have difficulty understanding premenstrual syndrome or don't want to be interested in understanding it. Premenstrual syndrome is merely the result of a hormonal deficiency.

Low thyroid symptoms caused by metabolic errors in a young boy's body or genetic inheritance factors will cause a physician to prescribe the thyroid hormone. The child may be placed on thyroid for the rest of his life. On the other hand, physicians are very reluctant to prescribe estrogen or progesterone for a woman with symptoms from low hormonal levels. Doctors resist prescribing hormones for women altogether. The premenstrual syndrome causes the world's most common major psychiatric illness. In fact, it has reached epidemic proportions because it affects over forty percent of all women at some time in their lives. Major psychiatric illnesses brought on by physical depletion of the hormone progesterone will cause the brain, as well as other parts of the body, to swell and retain too much water. This causes various physical and emotional symptoms, which occur regularly at the same phase of each menstrual cycle, that is, two weeks before the start of the period. These symptoms usually stop with the beginning of menstruation. The symptoms include tension, depression, anxiety, panic attacks, irritability, angry outbursts, crying without a reason, fatigue, forgetfulness, clumsiness, mental confusion, and craving for sweets, salty foods, or alcohol. Water retention may cause breast tenderness and the bloating of the brain or organs such as the stomach, ankles, feet, and fingers. Water retention causes joint pains, headaches, backaches, and susceptibility to genital herpes and sinus allergies since resistance is so low. The old saying goes, "hormones down, allergies up," therefore, sinus and asthmatic attacks occur more often during the two weeks before menstruation. Since the brain retains too much water, a person subject to seizures will suffer more seizures at this time of the month. Usually, a woman will have several symptoms. Often, she complains of feeling "out of control" physically or mentally. You have seen the waitress who dropped a piece of silverware every time she picked up the plates. A child can even sense these times of uneasiness from the body language of his mother. Menstrual cramps or sinus should not be confused with premenstrual syndrome. A depletion of the hormone progesterone causes symptoms before the onset of menstruation. History indicates that women suffered from temporary insanity, as physicians called it centuries ago. Queen Elizabeth was under the weather for approximately ten days out of every month. This probably was the reason that Sir Walter Francis Drake lost his head. This premenopausal syndrome begins in the early teens and worsens with age. By age thirty, it is recognizable, causing suffering, breakups of families, and requiring counseling for many persons. It is often called the ten-year divorce syndrome because if a woman has premenstrual syndrome when she marries in her twenties, it certainly will become worse, peaking around age thirty or forty. True menopause, or the depletion of estrogen and progesterone, usually takes place after age forty. Women are very influential when it comes to relating to their children, husbands, business associates, friends, and other people in business with whom she deals. She probably affects between sixteen and twenty-two other persons on average daily. When there is a depletion in any hormone, especially progesterone, water retention will affect the organs where progesterone receptors end. These recipients need progesterone to metabolize the water and nutrients correctly. When these endpoints are not satisfied, the whole metabolism of the organism goes awry. Too much water and salt retention will cause the symptoms mentioned. The brain also is affected by water retention. There will be a loss of potassium, which causes slowing down or short-circuiting of the brain's electrical impulses. In some women, premenstrual tension is alleviated by birth control pills.

A definite correlation has been found between women who miscarry frequently and who have a low circulating progesterone level. Dr. Katrina Dalton of London is considered to be the world's best-known authority on premenstrual syndrome (PMS). Dr. Dalton prescribed progesterone to women who had repeated miscarriages. After ten years, the patients who were kept on progesterone were not only doing well but also boasting about how well their children were doing in school. Inadvertently, we may have found that extra progesterone in a woman's circulating blood may, in fact, mean that she will give birth to a child with a higher-than-average IQ. M. Dorothea Kerr, M.D., Professor of Psychiatry at Cornell Medical School, finds hormonal replacement the medication of choice in patients with depressive states and emotional problems relating to the menstrual cycle. It has also been found by two other doctors, Dr. Yen and Dr. Reid of the University of California at San Diego, that even in severe cases of premenstrual tension and depression, one-third of the women will respond to estrogen and progesterone therapy. This group of one-third were, in fact, low in both estrogen and progesterone. Two or three cases in London where violent crimes have occurred have had the judges excuse the women as long as they sought correctional treatment from Dr. Katrina Dalton. These women were found to be temporarily insane due to severe progesterone deficiency. In its true form, Progesterone must arrive in the bloodstream without digestive changes. Progesterone, when given injectably or by vaginal suppository, goes directly through the walls of the vagina into the bloodstream unchanged. Interestingly, this procedure has been used in England for approximately thirty years. The United States has yet to pass permission for any manufacturing company to manufacture progesterone suppositories in America. So far, it is up to the enterprising physician and pharmacists to get together and talk the pharmacists into compounding suppositories by hand. Pharmacists cannot financially afford to make the suppositories by hand unless there is a great demand for them.

Some women can go without therapy for a period of three to nine months after receiving adequate progesterone replacement. There are relatively no known side effects to the use of progesterone.

Endometriosis

Endometriosis is often called the silent killer of the kidney. Endometriosis is a metabolic, hormonal disorder. Patches of endometrial tissue, or the inside lining of the uterus, are found outside the inner uterus. The uterus appears to spill its endometrial lining out through the tubes into the pelvis and surrounding areas. Patches of endometrial tissue go through the same changes as the menstrual cycle. In the first half of the cycle, the endometrial lining grows thicker. During the last half of the cycle, the endometrial tissue deteriorates and sheds, resulting in bleeding in the previous part of each cycle. Bleeding on the outside of the uterus or inside the pelvic area results in peritonitis, a severe inflammation of the smooth protective layer of each organ. Of course, this results in pain and edema. When the endometrial tissue is around the uterus or passageways from the kidneys, it results in the kidney, being unable to completely empty into the bladder. Rare cases of endometriosis have been found to spread throughout the body, including to the sinus cavities. This condition would cause hemorrhages in the sinus areas during the last half of every monthly cycle. Surgery or cauterization is needed to correct this condition. Endometrial tissue can cause adhesions of the ovaries, and they will appear to be less fertile. Adhesions in the pelvic area can partially or entirely obstruct the intestinal tract. Endometriosis is considered to be the silent killer of the kidneys. Pain in the pelvic area of the woman often hinders normal intercourse, which will cause emotional strain as well as self-doubts about the woman's sexuality. During pregnancy, a woman felt better because she did not have periods for nine months and also because her hormone levels of estrogen and progesterone increased, allowing a homeostasis condition whereby her hormones would be in balance again. The little patches of endometriosis on the outside of the uterus were put to sleep. Treatment for endometriosis is given in large doses of estrogen _and progesterone for a prolonged period of time, mimicking pregnancy.

Women under treatment for endometriosis often maintain their estrogen and progesterone at such high levels that they do not have periods indefinitely. Therefore, this treatment can replace surgery.

Pregnancy

Pregnancy brings about many complex hormonal changes. During pregnancy, a woman typically has a rise in estrogen and progesterone from three to six hundred times the average level. When this rise occurs, we find that endometriosis and rheumatoid arthritis are relieved. Formerly, these diseases have not been looked upon as hormonal deficiency except by a few physicians. Dr. Katrina Dalton of Great Britain knew that if women did not go through the expected rise in estrogen and progesterone hormones, they would suffer repeated miscarriages when pregnant. Also, an increase in blood pressure may cause kidney damage in the mother by allowing water retention from toxemia pregnancy. Hypertensive problems in the mother are also symptoms. Dr. Varron of Texas, former Head of the Endocrinology Department of the Baylor Medical School, has stated for many years that rheumatoid arthritis in the female is really a hormonal deficiency of estrogen and progesterone. The proof of this fact was that when one examines x-rays of the bones of many people with arthritis who have rheumatoid arthritis, osteoporosis can be detected. Dr. Varron has shown how to give hormonal injections of estrogen, progesterone, medroxyprogesterone, hydroxy progesterone, and HCG to men and women with rheumatoid arthritis. Approximately ninety-two percent respond to hormonal therapy. The American Arthritic Association informed Dr. Varron that he was on their political turf, but the tide has turned now. In 1982, Dr. Edwin Quilligan of the University of California at Irvine indicated that low serum progesterone results in twenty-eight percent premature births. When a pregnant mother whose hormones are two to six hundred times the normal level, she will possess a natural feeling of psychological well-being and strength. Women feel better naturally and have less pain and fewer physiological disturbances in their bodies when their hormones are at a high level during pregnancy. Women also have fewer allergies and infections during pregnancy.

One patient experienced a sudden positive personality change after becoming pregnant. Her complexion changed from off-white, sallow to that of a bright-eyed, pink-cheeked youngster. Her mood and sense of humor improved, her carriage improved, and her hair looked healthier. Even though she had many marital and family problems before pregnancy, she and her husband enjoyed a good relationship while she was pregnant. After the delivery of the baby, she was immediately put on birth control pills. She and her husband found to their delight, that her moods stayed level, probably due to the increased hormonal level from the birth control pills.

Another lady that comes to mind was twenty-three and suffered severe, debilitating rheumatoid arthritis. It seemed as if every joint in her body was swollen and painful. Her hands, ankles, shoulders, knees, and wrists were all swollen. She had rheumatoid nodules on her wrists and ankles. She became pregnant, and those nodules started to disappear in twelve to fourteen weeks. Her joint swelling went down, pain decreased, and, of course, she became a much more cheerful person. After she delivered, she had an acute flareup of rheumatoid arthritis and remained in pain with her arthritis for two or three weeks. We then started her on high doses of hormones, and she has successfully remained without rheumatoid arthritis since that time.

Postpartum Depression

Postpartum depression is a condition that can be very devastating. It affects the bonding that should take place between the mother and child, as well as the relationship with the entire family. In American Family Practice, August of 1982, Steven R. Lee, M.D., at the University of Southern Carolina School of Medicine, stated that up to fifty percent of all postpartum females have emotional dysfunction, which interferes with the close relationship with the baby and also has a devastating effect on the rest of the family. During pregnancy, the estrogen and progesterone accelerate two hundred to six-hundred times normal. Still, this roller coaster continues after delivery with a precipitous fall to the place where the estrogen and progesterone go below the typical values. As stated in the journal called The Female Patient, October 1982, an article called "Hormones and Depressive Illnesses" by Conrad Swartz, M.D., Chicago Medical School, it states that postpartum depression and premenstrual affective syndrome are both linked to a lowered estrogen level. A lack of thyroid activity often causes major depression. This shows the combined effect of what we call the "twin sister" effect; when the thyroid is increased, it increases the effectiveness of estrogen and/or progesterone. Postpartum depression often lasts from three to six months after delivery.

Another problem that might arise is a natural depletion of hormones after delivery because pregnancies often occur too close together. Generally speaking, if a woman has two children that are born closer than fourteen to sixteen months, that woman stands a very good chance of being depleted in one or all three of the hormones, estrogen, progesterone, and thyroid. Women cannot lose weight after pregnancy because their thyroid has remained exhausted and probably will continue to remain exhausted unless she receives thyroid pills. Combined with low hormones, a woman is now very subject to becoming sexually inadequate and depressed. Doctors can prescribe hormonal support that woman needs, and she can remain on a physiological and a psychological high from that point on, thus avoiding the terrible "baby blues."

Hysterectomy

Women often come to my office a few months after having a

hysterectomy. They do not understand why they can't quit crying, why they've lost interest in sex and why their husbands feel rejected. The children get on their nerves, their hair becomes brittle, their skin dries, and life just seems to be dull. They believe they have adequate ovarian function because their surgeon assured them, he left both ovaries. Let's look at the anatomy of what happens during a hysterectomy.

The artery that carries eighty-five percent of the blood supply to the ovary is suddenly removed during a hysterectomy. Menopausal depression follows. The small artery that leads into the ovary supplying about fifteen percent of its blood supply, comes from the pelvis rim. I recall that one of my professors in medical school stated that even if a woman retains her ovaries, she will need to be treated endocrinologically the same as if she had gone through menopause. Hysterectomies result in the sudden depletion of hormones, estrogen, and progesterone, which results in menopause. Other surgeries also place women at risk of losing part of all of their ovarian function. The butterfly surgery tubal ligation is one that may affect a woman. If careful dissection of the fallopian tubes were to take place, there might be little or no effect. However, usually, the surgeon simply clamps the whole area, resulting in the severance of eighty-five percent of the blood supply to the ovary. This woman won't be able to understand the change in her physiology eighteen months later. A few women with tubal ligations still have a sufficient hormonal level to keep their cycles but not enough hormonal supply to keep them at an optimum psychologically and physically.

Other surgeries that may cause interference with the blood supply to the ovary, of course, include the C-section or any abdominal or pelvic surgery that may cause a shocking event or trauma to the ovarian blood supply. Several months ago, a man brought his wife to our clinic. She was depressed, in severe pain, and could not get out of bed for three years. The husband had taken her to several different doctors, but they couldn't relieve her pain. I asked him, "Has your wife had a hysterectomy?" "Yes, she had a hysterectomy about twelve years ago. She has not been able to get out of bed for the last three years. Friends who come to the house look at her and say, 'My goodness, you're shrinking.'" The diagnosis was clear.

The woman had depleted her hormones so fast that the calcium in her bones was almost depleted. She has osteoporosis, thinning of the bone, and the collapse of a vertebra in her back which is causing "spinal arthritic pain." Her inability to move around without severe pain is caused by the collapse of the vertebra themselves in a compression fracture. This is much like an accordion. The spaces between the vertebra become narrow. The bones actually press on the nerves. Hormonal depletion secondary to the hysterectomy needed to be corrected by placing her on high doses of calcium with vitamin D. Since the patient was living in the mid-west, I advised them to find a doctor to give her hormonal replacement therapy immediately. The only doctor they found refused to give her hormones saying, "Don't you know that hormones cause cancer?" They were disappointed. They flew to California for this costly treatment which should have been available in their own hometown.

Most doctors warn women by indicating that hormones increase their risk of cancer from five to fifteen percent. What a terrible sentence to be imposed upon human beings. There is no turning back except through hormonal replacement or endocrine therapy. By giving a woman this life spring therapy, the physician affords her the chance to turn back the clock of time and keep herself physically and emotionally in better balance than she was before the hysterectomy.

Menopause

Menopause is a word women fear and misunderstand. Physicians would just as soon not hear the word because they believe there is "no cure." French physicians in the 17th century described a woman in menopause as "decaying from within." A woman will actually shrink in stature and lose her bone density and strength when deficient in the level of the sex hormones: estrogen and progesterone. Like hypothyroidism, menopause can occur at any age and for various reasons.

Let's, for a minute, look at the menopausal risks. First of all, it is the loss of loved ones. Not only does the woman/mother image in this condition become an irritable woman that John from college doesn't like to bring his friends home to meet, but she also has a lot of fears that are very subtle, to begin with, and then become tragic later. She fears her loss of sexuality. She may subconsciously fear or resent any or all other young women. Often mothers and daughters clash during this time in their lives. Divorces usually occur around the 25th wedding anniversary because many women at the menopausal age find they cannot cope with the symptoms that hormone deficiencies have brought on. These problems, of course, are interpreted by the husband as rejection.

The entire metabolism of the brain cells and their ability to react quickly depends upon estrogen and progesterone for proper functioning. Many times, the brain seems to short out, thus causing character disorders, personality changes, and forgetfulness. Along with this, a lack of hormones has a very direct effect on the skeletal framework. Without the proper level of hormones, the bones become thin, which is a condition called osteoporosis.

Pain, fractures, and deformity haunt a woman if this condition is left uncorrected. The decrease in energy, stature, muscle tone, and bone softening are demoralizing. She faintly remembers that smooth gait and stride that made her appear feminine, sensual, and attractive. Loss of energy is also the result of a lack of estrogen and progesterone. When the thyroid level drops, there is even a further lowering of energy. The risks in menopause are very numerous. Menopause carries with it the risk of increased heart attacks, hip fractures, depression, memory loss, senility, vaginal atrophy, and thinning of the muscles around the bladder.

Comparing a woman's menopausal risks to a woman receiving estrogen and progesterone is dramatic. The menopausal woman at risk, who is not receiving hormones, has three times the death rate from fatal heart attacks. She has five times the crippling strokes and eight times the incidence of cancer of the uterus, breast, and ovaries. She has fifteen times the incidence of rheumatoid arthritis and other conditions attributable to osteoporosis. She has two hundred times more episodes of depression and two thousand times more incidences of fatal hip fractures. One of the major thrusts of medicine today is trying to educate physicians that women do not have to be torn apart, crippled, maimed, and killed by menopause and post-menopausal osteoporosis. In America alone, currently, there are over 52,000 deaths per year from hip fractures due to postmenopausal osteoporosis. 85,000 women a year who survive hip fractures become senile because the fat globules that were released into their bloodstream from the bone fracture went to the brain. The bones are weak and chalky, and the hardware won't hold, so the bone cannot grow together again. 85,000 women are placed in convalescent hospitals annually. These senile women sit suffering results from postmenopausal osteoporotic fractures of the hip. Replacement hormonal therapy and calcium with vitamin D would have allowed these women to be restored to normal. Have you known a lady who was in her prime, youthful and energetic, and who is now reduced to an irritable person with insomnia, loss of sexual interest, extreme depression, fatigue, loss of youthful appearance, dry hair, dry skin, and a myriad of psychological and skeletal pains?

Women in the United States have very few options when it comes to menopause. Physicians either will not discuss it or may merely agree that it exists. Those who agree that menopause has true physical and psychological risks warn women concerning hormones and the increased risk of cancer. This is a totally erroneous statement. Physicians are stereotyped in their belief that menopause may exist only in the physical form of causing osteoporosis, drying of the skin, and the vagina. They generally do not believe menopause has anything to do with mental attitude, sexuality, depression, insomnia, migraine headaches, etc. Dr. Robert B. Greenblatt was indeed brave to write an article bringing to doctors' attention the many non-menopausal symptoms that can be explained and alleviated if one extends the definition of menopausal syndrome. He discussed hot flashes, the atrophic vagina, painful intercourse, infections, and bleeding. He continued by saying that this also ushers in fatigue, insomnia, depression, and anxiety, as well as sexual dysfunction and migraine-type headaches. Women respond to hormonal therapy and quit suffering Iron urinary incontinence and repeated bladder infections. Often records indicate that doctors termed urinary infections as sterile infections. Any slight agitation brought on symptoms mimicking urinary tract infections to the sensitive hormone-starved tissues. These symptoms are not relieved by the usual antibiotic therapy or dilation. Dr. Greenblatt states that the normal estrogen dose may be inadequate, and the oral therapy may not prove completely satisfactory; therefore, adequate amounts of hormones should be prescribed to correct the symptoms. In fact, progesterone is a great protection against the occurrence of cancer of the uterus, breast, and ovaries.

During menopause, it is not unusual to occasionally give a

woman with fatigue, headache, and frigidity testosterone, as we know that the mid-cycle ovaries normally produce testosterone to increase the libido. Nature increases the sex drive mid-cycle in women by producing testosterone, so she will get pregnant at this time. Women who are prone to headaches, both after menopause and also as a part of premenstrual tension, are easily relieved when given the correct hormones. If a woman is not relieved of headaches by estrogen and progesterone, it becomes clear in the medical literature that this woman would also need testosterone. The depletion of testosterone in the female usually results in headaches, fatigue, and frigidity. More than half a million women become of menopausal age every year. Most women who have lost their sex drive and may even be considered frigid respond very well to estrogen and testosterone therapy. This hormone can be prescribed orally, by injection, or by subcutaneous pellets. The richness of the life of any woman directly depends upon the level of hormones in her bloodstream. Many women suffer mood changes and psychological changes because their hormones are too low and require estrogen, progesterone, and/or testosterone. These hormones are psychotonic agents – which means that when given in pharmacological doses, they will have a direct effect on personality. These hormones directly affect the brain cells when given in higher-than-usual dosages. The only way to escalate this level is by injection or subcutaneous pellets. Many women do not respond when the medicines are given orally. Therefore, the blood level would never reach a high enough level to bring about any positive changes. Most women, because of the inconvenience of tablets and frequent side reactions caused by additional non-human hormones added to the tablets, choose to receive injections every two weeks. A frequent question asked by women is how long they will have to continue these injections. Endocrine therapy is needed only so long as the woman wants to feel good. The incidence of bone fracture, cancer, and fatal heart attacks, will return within 20 to 30 months after discontinuing any hormonal therapy. There is rarely a good reason for discontinuing hormonal therapy in a woman. Hormones should be looked upon as a restoration of the natural level of hormones in the bloodstream. It is the natural way to preserve life or to extend the prime lite. Women at the age of thirty to thirty-five should start on small doses of estrogen, progesterone, and possibly testosterone and thyroid if needed to keep their bodies at their natural prime state. This would avoid or bypass all of the unnecessary menopause and postmenopausal symptoms and tragedies. Preserving the healthy natural level of estrogen and progesterone should be considered with the same respect as the preservation of the natural state of insulin for a diabetic patient. The population of older women who look and act younger is now going back to college, traveling, working, entering the business world, and living fulfilled, prosperous lives, thanks to hormonal replacement therapy.

In summary, it is suggested that women take stock of how their hormonal levels may affect their body image. The following are indications of proper hormonal levels:

1. Proper hormonal levels assist in keeping weight down.
2. Sharp thinking and a keen memory are signs of adequate

hormones.

3. Good posture and poise, free from pain, stooping, or osteoporosis, are signs of sufficient hormones.
4. Hair is the crowning glory, healthy, thick, and alive when hormones are well balanced.
5. Breasts are firm and youthfully shaped.
6. Skin is tight and firm.
7. Warmth, sexual aliveness, and personality make a woman feel attractive and self-confident when hormones are present at natural levels.

A Rose That Doesn't Wilt

A rose without water soon is drooping. Often a rose bush can be brought back to life with watering, but often at the price of faded petals. The rose will be in better condition if kept in its prime in the full bloom state as long as possible. The same holds true for a woman who is aging rapidly after thirty-five and becoming elderly without the maintenance of the proper level of hormones in her body. Prevention is the answer. It's better to encourage a rose to live through spring, summer, and autumn years without fading or wilting.

Every time I go home to Connecticut to visit, I return to my mother's garden. Often late in the fall, I can observe the deep crimson brightness of a rose that seems to stand straight and strong long after other flowers have faded and wilted. Many times, colored oak leaves have fallen all around that rose. A woman is like that rose as she brings beauty, serenity, and strength to those around her. She takes in strength as though water for a rose, the stream of life is her hormones. She will add an extra dimension of color in spirit and zest to all those who know her.

Avoid Emotional Castration

The aura a woman exudes when she gets depleted of hormones is not fully known in medicine today. Women with high hormonal levels appear exciting to men, not by what they say, nor by how they say it, but by their well-being, poise, and confidence. Medical researchers do not understand the phenomena whereby a man is aroused by a woman. Whether there is a scent that is undetectable by smell, similar to animals being in heat, or if there is some ethereal aura or magnetic force which acupuncturists believe exists in the human body. These unknowns may be immeasurable; however, the effects and results are, in fact, measurable.

At a famous sex clinic in the United States, thirty-two executives were being studied regarding their impotence. The participants had been impotent for many months and had not responded to therapy. When reviewing their history, it was found that each of these men was married, and their wives were in their late forties. The study on these men paired them into two groups. One group of sixteen men was sent home with the arrangement that each of their wives would report to the clinic once a week for an injection of estrogen and progesterone. The other group of husbands with their wives was also given injections each week. The husbands in the second group received colored water, and the wives received estrogen and progesterone. At the end of twelve weeks, the statistics were reviewed. There was no statistical variance in either group. Both groups responded by having the majority of men regaining their sexuality. None of the men were treated by counseling or hormones. When the women were treated with hormone replacement therapy, positive changes took place in the area of sexual relations. Even though it cannot be understood or analyzed scientifically, it has been documented by controlled studies.

The danger of burnout, career failures, and health problems also become factors to deal with during menopause. Every day there are many examples of the executive woman who has done very well in her career and suddenly takes a devastating plunge into depression, irritability, drug abuse, alcoholism, or poor health. These problems are essentially correctable through endocrine therapy.

Many doctors seem to refuse to face the psychological issue of menopause because it is far-reaching and affects so many people. Children and teenagers are directly affected when their mother suffers menopausal symptoms. It is difficult to understand family problems when the mother is depressed. Could it be due to difficulties the teenagers are having with drugs or in school or is the reason the young people are doing poorly in school or on drugs because the mother was depressed?

One lady came to me after twelve years of being in and out of institutions and having had much psychiatric counseling. She feared driving or even leaving her bedroom and could not collect her thoughts or strength to do her housework. She had been unable to assist in family planning or cooking for twelve years. She came for help because of her severe, unremitting depression. She also was suffering a tremendous amount of spinal arthritic pain. Her family and physicians were unable to understand her depression. I placed her on endocrine therapy, and within one week, she showed improvement. Increased doses of endocrine therapy were prescribed for eight to twelve weeks. This therapy occurred just before Christmas, which had always been a traumatic time in her life. Depressed people always seem to be more depressed over the holidays. Between Christmas and New Year's Day, her family came to my office. The husband spoke, saying, "Dr. Corley, I want to thank you for the return of my wife and my girlfriend – if you know what I mean." The daughter came over, kissed me, and said, "Do you realize what it's like, Doctor, to have your mother not only share Christmas dinner with the family for the first time in twelve years but to have actually prepared it herself." What a wonderful Christmas present this family had experienced. Many families may miss a mother, a wife, and a lover because some doctors refused hormonal replacement therapy at the right time.

Osteoporosis

Physicians for years have felt that parathormone, which acts as an agent to drain calcium from the bones, has nothing to do with osteoporosis. Yet, we now believe that estrogen itself is the counteractant agent to parathormone. The function of pharathormone is to help raise the calcium level in the bloodstream. Therefore, it robs calcium from the bones. Estrogen has been found to act as a protection for the kidney against premature aging. When the kidney ages due to a lack of estrogen, it no longer has the same ability to convert Vitamin D to Vitamin D3, its active metabolite. Vitamin D3, the active metabolite, is needed to send calcium through the intestinal wall into the bloodstream so that parathormone can use it to keep the calcium levels up in the bloodstream.

Estrogen protects the bones so that it seems to guide the parathormone so that it will take the calcium it needs from the intestine and from the atherosclerotic plaques in the walls of the blood vessels rather than from the bone itself.

Rheumatoid arthritis is a type of osteoporosis that affects the elderly in our population. We refer to the increasing numbers of people in their sixties and seventies as the "gray boom." Osteoporosis is known as the killer epidemic. It is the leading cause of crippling and death in women. Over 15 million people in the United States currently have osteoporosis. Over four million are in severe pain. Over one million people annually suffer from osteoporotic fractures. Fifty-two thousand do not survive, and another eighty-five thousand become crippled or senile when the newly fractured bone sprays millions of fat bubbles into the bloodstream, which move within the bloodstream until they settle in the brain. Menopause and hip fractures are the most crucial dangers among aging women. Hip fractures usually require surgical pinning, which often is unsuccessful because the bones are too thin and fragile. Hip fractures are the twelfth leading cause of death.

More women can now enjoy a more pleasant life if they receive endocrine therapy, a good diet, moderate exercise, and calcium with Vitamin D supplements. This is a relatively easy formula to follow to prevent such severe conditions from affecting many people. Unfortunately, the costs of treating the aged or geriatric medicine are continually rising. Since this condition is almost totally treatable and preventable, one wonders why any woman would refuse the benefits of endocrine therapy.

Dr. Gordon studied two hundred and twenty women who were already advanced with osteoporosis. He decided to place them on estrogen and progesterone therapy and study them for thirty years. Many of these women were already in their sixties. Many of these women had reduced incidence of heart attacks and strokes because of the estrogen and progesterone therapy. Even more important was the fact that none of these women suffered any hip fractures. Another study done by Burch, Bird, and Vaughn in 1975 involved over fourteen thousand patient-years of women on estrogen and progesterone. There were no incidents of hip fractures among this group either. When hormonal therapy is discontinued, it takes only two years for the same risk of osteoporosis, hip fractures, cancer incidence, heart attacks, and strokes to return. When hormone therapy is discontinued, bone depletion accelerates dramatically.

Obese women generally have an excess amount of estrogen due to a neutral hormone from the adrenal glands, which metabolizes androstenedione into a female hormone called estrogen. When estrogen is high, and out of balance with progesterone, these women are at a greater risk of cancer of the uterus. Progesterone assists in the formation of harder bones. These women should be kept on cyclic progesterones. Back pains that do not show on X-ray respond to endocrine therapy, mild exercise, and calcium with Vitamin D. X-ray diagnosis will not indicate osteoporosis until over fifty percent of the bone mass has been depleted. Many women who lose weight by extensive exercise or running have a low level of estrogen and progesterone either because the body fat is too low or because their expenditure of energy is used entirely for their athletic endeavors and not enough is left over for their endocrine development. Many of these women go without periods due to low hormonal levels. These women, who are healthy in every other respect, would generally be expected to have harder bones than usual because of the exercise factor. Due to hormone depletion instead, these women are at a greater risk of osteoporotic fractures than if they were not on extensive exercise programs.

A hysterectomy will cause premature menopause. After a hysterectomy, a woman can suffer a loss of five to twenty percent of her hormones each year. This fact was unknown until recently. Factors that accelerate a tendency toward osteoporosis are diabetes, a sedentary lifestyle, inactivity, taking of cortisone, no pregnancies (remember that during pregnancy, the hormone levels go up two to six hundred times which increases the bone mass), early menopause, due to low hormones throughout her life, little exposure to sunlight (this has to do with the conversion of Vitamin D to its active metabolite), and drinking caffeine. Caffeine uses up calcium excretions. Cigarette smoking, alcohol, and vitamin deficiencies are also contributing factors. Excessive amounts of Vitamin A or D or protein in the diet will also predispose osteoporosis conditions. Calcium deficiencies in the typical American woman's diet are much too prevalent. Women should take 1000 to 1500 mg of calcium daily. Less than 500 mg in the average American woman's diet is considered average. Women in their 20's and 30's are already showing signs of osteoporosis. Twenty-five percent of young women already have osteoporosis. Probably ninety percent over forty-five have osteoporosis. Osteoporosis should be treated preventively rather than waiting for tragic events of vertebral compression or hip fractures to occur. The dowager's hump in the back results from crush fractures of the vertebra. Estrogen, by itself, is protective against the resorption of bone. It also increases the bone protein matrix. Therefore, it is vital for therapy to start early. Progesterone aids in the hardness of bone and balances estrogen in the body physiologically. Physicians who prescribe the smallest dose for only a short duration are allowing women to risk heart disease, stroke, cancer, severe osteoporotic pain, disfiguring, and fractures. Black women, due to heavier skeletal bone masses, and also women who have fat density, have increased estrogen levels and, therefore, a lesser degree of osteoporosis. A recent study at Mayo Clinic indicates that there are over a million hip fractures a year. One-third of all women over sixty will have one or more hip fractures. The aging population of the

United States is increasing. The population over sixty will double in America within ten to fifteen years. One-fourth of all women over sixty already have osteoporotic fractures or vertebral crush fractures of the spine, which may not even have been diagnosed by X-ray as yet. These vertebral fractures are crippling. Estrogen and progesterone, androgen, and testosterone can be prescribed to reverse osteoporosis. The typical woman who gets severe osteoporosis pain is a slender, frail, aging woman who undergoes many psychological and physical changes. Dr. Norton of the Medical Research Council in Leeds, England, states that it is astonishing how many patients have difficulties in the sexual area of their lives. Hormonal therapy can correct hot flashes in two to three weeks, but sexual problems become deeply established and are not quickly resolved.

Dr. Paul J. Fink, MD, Professor, and Chairman of the Department of Psychiatry at Jefferson Medical College in Philadelphia, in an interview for the JOURNAL OF SEXUAL MEDICINE, says that no one is really paying attention to the diminished sexual interest and activity of osteoporosis patients. Altered body image and pain can change both the treatment and the prognosis of the treatment of depression as well as osteoporosis.

Women who are kept on estrogen are known to have one-third the death rate of women not taking this hormone. Physicians in the past who have made a plea for the maintenance of adequate estrogen from puberty to the grave have reported in articles that have appeared in The Journal Of The American Geriatric Society. Among those doctors are Drs. R. A. Wilson and T. A. Wilson, husband and wife; Dr. Albright, Dr. Greenblatt, Dr. Gordon, and many others. There have been reports from Dr. Atkin and Dr. Lindsay of Glasgow and Dr. Norton of Leeds, England. Dr. Greenblatt points out that conservative clinicians who argue that one should administer hormones only in small dosages for short periods are allowing these women to be exposed to bone loss and osteoporosis. Dr. Robert Heany of the American Society of Bone and Mineral Research from the University of Omaha, Nebraska, states that one-fourth of all females over sixty already have fractures of the spine. Humanitarianly and financially, this condition goes untreated or unchecked in this country. This massive epidemic is raging throughout the female population at an increasing rate. It must be stopped.

Dr. Patrick W. Irvine of the University of Minnesota Medical School says that over fifteen million patients have osteoporosis today. Four million are suffering severe pain. Over one-third of all white females will have one or more hip fractures in their lifetime if they do not receive endocrine therapy. Hip fractures cause more suffering and deaths than cancer. The American way of life, which is a sedentary lifestyle as opposed to being the hard-working immigrants, increases the tendency toward osteoporosis. For every death caused by endometrial cancer, there are forty deaths caused by hip fractures for women who have not received estrogen. Therefore, even if estrogen therapy is given with poor medical supervision, it provides a woman with a better chance than had she not received any hormones at all.

Over 75,000 hip replacements and 40,000 knee replacements are performed annually in the United States due to osteoporosis. Therapy can be given even if cases are advanced and there will be some response. Lessoning of pain, even if it is too late to correct a deformity, is an improvement. Most women are remarkably active and able to cope with the usual activities of daily life until a hip fracture occurs. The medical/psycho-social outlook is very dim without hormonal therapy. Medicare costs for hip fractures in 1983 will pass two billion dollars. Many patients will remain severely incapacitated and unable to leave the convalescent hospital because they suffer a loss of independence, social contact, severe pain, confusion, and depression. This means a great loss to families and society.

An elderly lady who already showed X-ray evidence of osteoporosis and a couple of vertebral fractures of the spine was admitted to a hospital because of pain in 1970. She returned in 1979, again, with severe pain. This time she was found to have twelve vertebral compression fractures of the spine. She still was not placed on therapy and, within that year, was readmitted due to suffering a pelvic and hip fracture. After surgery, she suffered from cardiac and pulmonary complications typical of osteoporosis patients. Her recovery was both lengthy and painful. The surgery was considered a success; however, she became confused and could not sit in a chair due to brain damage from the spray of fat emboli in the bloodstream when her hip fractured. Hormonal therapy was never given. After being transferred to a convalescent hospital, she only lived for two months. This tragedy was unnecessary. If she had received hormonal therapy, there would have been a different outcome to this story. X-ray evidence is always late; therefore, it is imperative that doctors treat osteoporosis preventively before a tragedy occurs. Once again, correctly given estrogen and progesterone will prevent osteoporosis and reduce the incidence of breast, ovarian and uterine cancer, atherosclerosis, heart attack, strokes, and hypertension and relieve most cases of rheumatoid arthritis. The only side effect that a patient can have while on this therapy is occasional menstrual periods at any age unless she has had a hysterectomy. This period will come during the progesterone part of the cycle. Menstruation is a relatively small price to pay for the prevention of tragedies that surround menopause osteoporosis. These women always look and feel younger and stronger.

Other guidelines given to the public about preventing osteoporosis include that behavior modification has to be started in childhood and continue through adulthood. A woman should take at least one thousand to one-thousand five-hundred mg. of calcium daily. One should receive adequate amounts of Vitamin A, Vitamin D, and protein but never excessive amounts. Vitamin A should be twenty-five to fifty-thousand. Vitamin D should average four to eight thousand per day. Brisk walking is also an excellent exercise. Cigarette smoking, alcohol, and caffeine should be avoided. The elderly should be safety conscious and prevent falls by avoiding high heels, removing throw rugs, avoiding polished floors, and installing adequate lighting. These are all significant issues because more than twenty-three million Americans are now reaching sixty-five or older. Over the next fifty years, this number will more than double.

Prevention has not been adequately emphasized because early osteoporosis is not usually visible. Young and middle-aged patients with risk factors must be identified and treated early. X-rays do not indicate the diagnosis until forty percent of the bone has been lost. Fractures occur when forty to fifty percent of the bone has been lost. So, the first sign of osteoporosis on the x-ray is usually the vertebral fractures which occur early in the disease. Once estrogen and progesterone therapy is started, the bone will grow stronger. After the use of hormones is discontinued for any reason, the deterioration of the bone is very rapid. The biological clock seems to hurry to catch up to the level of osteoporosis that the woman would have had if she had been untreated. This acceleration takes two years. The moral of the story is to never stop endocrine therapy once it has begun.

In a recent article, Dr. Judd estimated that there are over thirty- million females of menopausal age who are in danger because o1 a hysterectomy. That is an astounding statistic. It is important to realize that at menopause, there is a steep drop of hormones to the level of only five percent or 1/20th of its original level. At that level, hot flashes and bone destruction continue at a very rapid rate. Demineralization often takes place at the rate of five to twenty-five percent per year in the first two years after hysterectomy or menopause. The medical doctors who support the use of hormones in the treatment of osteoporosis declare there is more suffering and deaths from osteoporosis than from endometrial cancer. As mentioned earlier, there are over 52,000 deaths from hip fractures alone, whereas only approximately 3,000 deaths from endometrial cancer. The amount of personal suffering and pain that occurs because of osteoporosis is probably incalculable.

Hopefully, in the future, the cause of the condition that we now call Rheumatoid Arthritis will be called hormonal depletion. Therapy will be prescribed as outlined by Dr. Harold H. Varon, M.D., Past Director of Endocrine Research at Baylor University in Texas. Dr. Varon published his results, showing a cure rate of approximately ninety-two percent for rheumatoid arthritis in females and males using sex hormones. The treatment not only resulted in the obvious cure for arthritis but also increased the energy, sex drive, and youthful appearance of these patients. Over fifteen million Americans have arthritis presently, and this will be a major breakthrough for those persons. Some patients of Dr. Varon, who had suffered for over forty-seven years, achieved relief from pain. One of the things that were pointed out about rheumatoid arthritis when I went to medical school was the characteristic that x-rays often indicated osteoporosis conditions. Why osteoporosis was present in the condition called rheumatoid arthritis was not known then. Dr. Varon's idea of treating arthritis, especially rheumatoid arthritis, with sex hormones will become a prolonged overdue medical therapy.

Rheumatoid arthritis falls into the same category in many respects as post-menopausal osteoporosis. Dr. Varon of Baylor Medical School of Texas made the statement that the disease, rheumatoid arthritis, was really a misnomer as it was just another form of post-menopausal osteoporosis. The proof he felt was when ninety-two percent of the people with rheumatoid arthritis, when given sex hormones, responded and were relieved of pain, symptoms, and occasionally even the deformity of rheumatoid arthritis. Opposers argue that rheumatoid arthritis can begin at a very early age; however, we have discovered that menopause and osteoporosis are caused by a hormonal deficiency and is not a respecter of age either. A baby can be born with a hormonal deficiency. Regardless of the onset of this "rheumatoid arthritis" or a specific type of osteoporosis, it is due to hormonal deficiencies. The crippling changes that take place in the bones can be reversed with the introduction of high-dosage hormonal replacement therapy. Another reason opposers did not believe that rheumatoid arthritis belonged in the category of post-menopausal osteoporosis was that the American Arthritic Association and the Local and County Board of Medical Quality Bureaucrats decided to discharge Dr. Varon from his position as Head Professor of Endocrinology. They also began a series of harassment incidents and lawsuits, which could possibly extract Dr. Vernon's medical license. Fortunately for Dr. Varon, the ongoing Twenty Year Study being done by the National Institute of Health gave strong indications that when women took the hormones estrogen and progesterone, amazing results were apparent. Rheumatoid arthritis is less than ten percent of that age group's expected incidence. It is interesting to note that Dr. Varon had been publishing throughout those same twenty years that between ninety and ninety-two percent of the women were cured of rheumatoid arthritis when they received hormones. Thus, to the embarrassment of many people, the Twenty Year Studies of the National Institute of Health verified Dr. Varon's position, and then the shoe was on the other foot.

Severe crippling deformity, human suffering, fatalities, and medical costs, which currently run between three and four billion dollars from osteoporosis and rheumatoid arthritis, are entirely preventable. In addition, senior citizens in greater percentages of the population are causing a "Grey Boom" in the United States. Over the next fifty years, the number of people over sixty-five will double, and the number of people over age seventy-five will increase five times. It is high time for the medical profession to be ready for these statistical increases in disease caused by hormonal deficiencies or else to plan to prevent these diseases by massive re-education of all physicians.

When current medical students reach their prime in their careers, at least seventy-five percent of their time will be spent dealing with the problems of the elderly. The "Grey Boom" will last longer than the "Baby Boom" of the fifties and will require attitudinal changes among family physicians. The proportion of the population that is over sixty-five is currently ten and one-half percent. This percentage is expected to increase rapidly over the next fifty years. The elderly are now better educated, enjoying better jobs and benefits, and are generally more financially secure. They definitely have higher health expectations. The family physician will need to understand more about the aging process.

It is becoming increasingly important to advise people regarding proper health habits and maintenance so they can enjoy life to the fullest. Health in old age is usually a reward for persons who have been healthy at a younger age.

A moderate exercise program throughout a lifetime will keep the basal metabolism from slowing down rapidly. This slowing down process would have a harmful effect on the brain. Mental confusion, loss of mobility, and inability to cope with stress are often not related to disease but to brain decay, which is often the result of inactivity.

Chapter 6: Menopause In Men

Menopause is no respecter of age, whether female or male. Babies who fail to thrive, young children who fail to grow and develop, children between the ages of six and twelve who suddenly do not gain weight and become temperamental, and teenagers who suddenly become erratic are all examples of low hormonal levels in children. They can easily be diagnosed and treated by giving testosterone which they are lacking. Sometimes HCG, which is Human Chorionic Gonadotropin, is lacking, which is needed to stimulate the adrenal glands to release a host of other sex hormones. The reason that we gloss over male menopause, which starts at about age thirty-two to thirty-six and continues for the remainder of life, is that it is a gradual decline in hormones that is usually not recognized. Dramatic changes occur in women's lives when they have a steep decline in hormone levels. The male hormone does not give protection against cardiovascular disease, so there is no sudden risk. More noticeable in the male population is the fact there are not many male athletes after the age of thirty-two. When the hormones decline, outstanding athletes start losing their reaction time. As was true with the twin sisters, when the sex hormones reduce, the thyroid function also decreases. Muscular endurance and energy become less. Testosterone, the sex hormone, is responsible for a person's endurance and energy level. Healing takes longer, too. So, we find fewer athletes and less robust behavior in men over age forty.

Other changes also take place, which is sometimes mistaken

for male menopause. Burnout of executives, primarily emotional and psychological, is a different symptom. Thinking in the cerebral portion of the brain and the hypothalamus tracks that end up in the pituitary give the pituitary a dull dark brown coloration or feeling. The pituitary reacts as if it is supposed to match dullness with dullness. Thinking negatively or experiencing negative feelings can, in fact, throw a man or a woman into menopause at any age. Negative thinking can cause hormone levels to decrease, and low levels of hormones can cause the thinking to become depressed. An old saying, "What you think, you become," is undoubtedly true. Male menopause is often the counterfeit term or the "cop out" term for various other symptoms. One of the first things that wives often come in and talk to me about is they feel their husband needs hormone shots because he hasn't been able to perform sexually for many years. It was a gradual thing and was probably caused by a number of other factors, such as taking each other for granted. Often couples are not innovative or creative in their marriage. Romance or personal interest and sex have become a thing of the past. Living with each other and eating a meal together occasionally becomes boring. Becoming less motivated, other things take priority, and the husband often becomes relatively impotent. His problem would have nothing to do with his hormone level. At least not at the beginning; however, what you think, you become, so secondarily he may have low testosterone from depression, but the treatment is not to give testosterone injections. Psychological counseling in which attention is drawn to the fact that his problem may be in his priorities in life, fear, depression, or burnout. These problems have nothing to do with hormones. Occasionally a physician will give triple doses of testosterone three times a week for two weeks, and then when there is no improvement, the patient will say, "you were right, doctor. I'm willing to think of other ways of approaching my problem."

Male menopause impotence is the last symptom that will show up if the hormone level is low, and even then, that is usually after the male reaches seventy or eighty years of age. Impotence at any age, besides being caused by emotional trauma or boredon, can be caused by diabetes mellitus. Sugar diabetes tends to cause peripheral neuritis, which may or may not be felt as pain. This type of diabetes seems to cause a fraying of the marlan sheath, which is a saran wrap-like film that surrounds our nerves. The sugar content and other toxins of sugar diabetes actually attack the nerve itself. One of the most vulnerable nerves of this type of circulation is the peroneal nerve. This nerve goes to the vasal constructing muscle, which clamps down around the dorsal vein in the penis and prevents the blood from backing out of the penis when the blood is pumped into the penis. This overflow of blood will fill the side and spongy areas and cause a hardening or an erection. This nerve may become damaged and then will not contract that muscle. The result is impotency.

Alcoholism will also bring about the same chain reaction causing impotency. This problem can be corrected by surgery when a small mechanical pump can be implanted in the abdomen, which will press down around this vein and bring about an erection.

Male menopause can be compared to female menopause. A sudden decrease in male hormones, or testosterone, will bring about fatigue, headaches, and a lack of sex drive. Those are the three outstanding symptoms. When a woman has those same three symptoms, she may also lack testosterone. Remember, women also produce a small amount of testosterone mid-cycle every month. Testosterone is one of the critical hormones which is responsible for keeping the bone hard in the female and the male. Progesterone, the female hormone, also does an excellent job of keeping the bone hard for women. She would really not need to be given testosterone to strengthen the bones unless her bones are crumbling when her condition is discovered, and the physician must act quickly. In that case, large doses of testosterone and progesterone will need to be administered. Testosterone drops very slowly in the male, and it would be fifteen to twenty years later in life when he becomes osteoporotic. Sixty-year-old women are prone to hip fractures if they have not been given hormonal replacement therapy, whereas the male is around seventy-six to seventy-eight before they would run into the same problem. The depression factor in men does not parallel that of women in showing up with sudden personality changes. Depression in men takes place over ten or fifteen years and is often attributed to the frustrating effects of aging, lack of control over one's own destiny, lack of money, or other health crises due to old age.

Headaches, depression, fatigue, and sometimes impotence due to burnout can begin with the thinking process in the emotional center, called the thalamus or hypothalamus of the brain. The brain gives signals to the pituitary indicating everything is dull. The pituitary gland, in turn, gives out hormones from its control panel that make the body react in a dull way. This reaction suppresses the testosterone output in the gonads of the man. The recommended treatment is a change of lifestyle brought about in conjunction with counseling and, lastly, the administration of testosterone injections if still indicated.

Chapter 7: Reversal Of Aging

Sexual Regeneration

Premature aging can be averted by guaranteeing an optimum hormonal level and thus enhancing the quantity and quality of life. The National Twenty Year Studies reported that when the correct amount of estrogen and progesterone are maintained, the following statistics were documented.

The group of women who did not receive hormones, statistically incurred three times the death rate from heart attack, five times from paralyzing strokes, eight times as much cancer of the breast, uterus and ovary, over ten times as much rheumatoid arthritis and vertebral fractures, twenty times as much severe depression and two-thousand times as many hip fractures. Greenblat and other authors brought out the fact that in giving pharmacological or larger-than-usual doses, hormones are, in fact, the best natural tranquilizer and energizer known to mankind, due to their direct and positive effect on brain tissues.

Bypassing Menopause

Bypassing menopause is more than just sustaining normal hormonal levels. Women who are subject to uterine cancer can experience an arrest or even a reversal in the cancer of the uterus if cancer is already present and has not progressed beyond the walls of the uterus. Medical reports which followed fifty-two women with cancer of the uterus for ten years found all cancers reversed and became benign. Four reversed back to hyperplasia, which is but a benign overgrowth of the uterine lining. The forty-eight that remained were reversed entirely to a normal endometrium with no trace of cancer. In the following ten years, while kept on progesterone, these women enjoyed freedom from cancer.

Physical, social, emotional, and mental health will also directly affect the brain and muscle tissues, as well as influence bone growth and the proper synthesis of proteins in the body when hormones are sustained. A depletion of hormones, therefore, results in muscle atrophy, osteoporosis, and brain atrophy which causes senility.

Counseling is of no value until the hormone depletion is corrected, which can bring about a normal sense of well-being and self-worth. Studies from the sixties, which erroneously and supposedly "proved that estrogen caused cancer," were extremely harmful to the women who were disallowed proper treatment of their hormone deficiencies. Three times as many women die every five years after the age of sixty due to strokes and heart attacks than if they were taking hormones. Added to those facts are the statistics regarding fractures and pain in the spine and the whole skeletal body. Needless suffering among women has been allowed due to the misunderstanding regarding hormones and hormonal replacement therapy.

Women who have taken birth control pills for five years or longer have been shown to have an incidence of breast and uterine cancer only one-half to one-eighth as often as women who did not take birth control pills. Birth control pills are made from estrogen and progesterone. After the age of thirty-two or thirty-six, when the hormone levels start to decline, these women will become susceptible to cancer if they do not receive hormonal replacement therapy. Fatal coronaries in women who do not take hormones are three times that of women who take hormones. In most cases, the quantity and quality of life seem to be simply a matter of hormonal replacement.

Sex desire, or libido, is attributed to yet another male hormone called androgen (or testosterone). It must be remembered that women produce predominately female hormones; however, they also create an appropriate, even though significantly less amount of necessary male hormones. Often the male hormone in women will disappear long before the female hormones get noticeably low. The exact reverse is true for males. Dr. Greenblat estimates that between six and eight percent of all women of any age lacking in libido are lacking androgen, which can also be restored by including androgen in their hormonal therapy. Caution should be exercised regarding the administration of male androgen to women. A minimal amount should be prescribed, and the patient should be watched regarding any male symptoms of lowered voice or growth of hair on the face. The alert physician will adjust the prescription according to the actual needs of the woman.

Men similarly begin to lose their hormones at age thirty-two or thirty-six. However, they do not experience a sudden drop in hormonal levels. The male has a very gradual decline in his hormonal level. He retains his normal hormonal level for about ten years longer than his woman counterpart. Brain atrophy can begin around forty-five in women and about fifty-five in men if hormone levels drop dramatically. Hormonal depletion begins around age thirty-five in women. Ten years with insufficient hormones will cause the brain to start to atrophy, shrink, and shrivel up. One of the first symptoms of atrophy of the brain is forgetfulness. Hormonal restoration will cause the memory to return in most cases.

Bone loss in women is greatest immediately following menopause. The relative risk of developing osteoporosis is also influenced by hereditary factors, which are not amenable to behavior modification. The major dietary or behavioral factors could be changed with sufficient motivation. For example, any factor that decreases calcium and Vitamin D ingestion or increases urinary calcium excretion will aggravate osteoporosis. Many elderly persons lack protein and vitamins in their diets. Caffeine and nicotine usually cause the acceleration of osteoporosis. Every cup of coffee destroys five mg. of dietary calcium. Coffee drinkers usually do not drink mild, therefore, they may have a lower intake of calcium. Poor nutrition and higher incidents of falling, and loss of calcium from caffeine, nicotine, or alcohol place the elderly in jeopardy.

Osteoporosis is believed to worsen with inactivity. Even minimal movement, such as rocking in a rocking chair, would allow exercise to promote the strengthening of the bones. Conditions that stimulate good bone formation are: (1) Adequate sex hormones of estrogen, progesterone, and testosterone as needed; (2) Adequate dietary intake of calcium, over fifteen-hundred mg per day, either in the form of milk, foods, and/or calcium supplements which must also include about four-hundred to eight-hundred units of Vitamin D daily; (3) Exercise several times a week in the form of a brisk walk for one or two miles or other enjoyable physical exercises; (4) Sun exposure of ten to fifteen minutes per day for four days a week would allow the ultraviolet action on the skin to convert Vitamin D to its active metabolite form.

Chapter 8: Health, Happiness, And A Well-Balanced Personality

Controlling Your Own Destiny

You can be in control of your own destiny when you realize certain factors. Number one, you must maintain a normal hormonal level throughout your lifetime. Secondly, positive thoughts and attitudes which cause reactions of the hypothalamus-pituitary chain to increase the output of your hormones should become a natural assurance of hormonal production. Thirdly, hormones given by injections or tablets need to be requested from your doctor.

Choosing The Right Doctor

Choosing a doctor who is knowledgeable in prescribing hormones to maintain normal hormonal levels may be somewhat tricky. The mainstream of medicine at this point, and for perhaps the next ten years, may not progress far enough to agree with the factors, decisions, and successes that we attribute to hormonal replacement therapy. First, call the physician's office and ask to speak to the head nurse. Ask three questions before making any appointment. The three questions are as follows: Does your doctor believe in and routinely prescribe hormone replacement therapy? Does your doctor give estrogen tablets as well as estrogen injections? Does your doctor give progesterone tablets as well as injections? If you receive yes to all three questions, then make an appointment to see the doctor.

Another important question that may be difficult to ascertain is regarding the frequency with which patients are treated by this doctor for hormonal replacement therapy. If the nurse suggests that the doctor rarely treats a patient for hormonal replacement therapy, it would appear that hormonal replacement therapy would not truly fit with that doctor's personal philosophy. After making the appointment to see the physician, then continue to do your homework on a full sheet of paper, print or type, double-spaced, a list of your physical, emotional, and mental symptoms. Be as organized and efficient as possible, using only a few words to summarize your complaints or symptoms. If handled properly, this procedure will save a lot of time. Otherwise, the physician must quiz the patient until finding and extracting the true complaints and symptoms. In addition to the symptoms you have listed, you may also write that you believe you need hormone replacement therapy according to the reasons you have read or learned.

Most doctors consider that any cancer in your family will significantly increase your risk of cancer and would be a direct contraindication of hormonal therapy. Therefore, you must convince the doctor of the strength of your convictions toward hormonal replacement therapy. I have had patients respond by saying, "I know, I had an aunt that died of breast cancer, and I understand that increases my risk towards breast cancer. My life, however, is in shambles the way it is now. I choose to take that risk by having hormonal therapy." Continue by indicating that you have read that prolonged estrogen therapy actually causes a decrease in breast cancer in most women. Most breast cancer incidentally is estrogen nondependent; therefore, cancer grows when there is a lack of estrogen.

Conversely, the cancer is restrained by introducing estrogen. About one-third of breast cancers grow faster when given estrogen. Thrombophlebitis, either due to birth control pills or any other factors, gall bladder trouble, or cardiovascular problems in your health history, may make it contraindicated to give hormones. Tell the doctor the facts. Be utterly truthful regarding your health history. Do not cause the doctor to feel paranoid in that you consider him medically and legally responsible rather than being the healing physician. Act calmly and without malice. Doctors who are knowledgeable about hormones will want to help you toward health and happiness in your life.

If you cannot locate a local physician who can prescribe hormone replacement therapy, you will need to consider an alternative. First, seek out a doctor who qualifies, regardless of the location, and have him draw up a prescription for hormonal replacement therapy, injections, tablets, and method of treatment which will be required, and request that he write that prescription for your therapy to take back to your local physician. If in a rare instance, you cannot find a physician in your area, you should call the nurses directory for a qualified registered nurse who can follow the prescriptions to complete your therapy.

Vitamins, Nutrition, And Exercise

Hormones cannot replace good nutrition, vitamins, exercise, and a correct lifestyle. Hormones work hand in hand with good nutritional practices. The feeling of "well-being" and "bend in control" will start to deteriorate quickly whenever the foundation of hormones begins to decline. For maximum health, a normal balance of hormones should be maintained along with good nutrition, vitamins, and plenty of exercise.

The Choice is Yours… Feminine Hormone Revolution

Imagery and visualization are really subjects that are written about in women's magazines in the United States today. These articles are read avidly and often followed as a cookbook approach. Sitting down, relaxing all muscles, closing the eyes, and visualizing a favorite or peaceful place has been found to be highly therapeutic. Further visualization of the white blood cells coming to the rescue for the purpose of healing and the pain as taking the form of an animal or a Disneyland character with whom you can talk and reason is of great benefit. These techniques are readily available in many bookstores. We become what we think; therefore, imagery and visualization, whether the person realizes it or not, can be utilized to form deep imprints upon the subconscious. In turn, the hypothalamic emotional brain centers will be affected, and then a positive influence will be felt on the pituitary, which can, in turn, begin a chain reaction of the healing process. The pituitary acts as the control panel, and our conscious thoughts, verbalizations, and even attitudes produce constant input to the pituitary. This input directly affects the level of hormones that are produced by the pituitary. Our bodies physically and physiologically become well-balanced in direct relation to the hormones produced. Happiness or the lack of happiness will generally result in parallel to our well-being or disastrous health problems, whichever the case may be. Psychosomatic or stress illnesses, which result in heart attacks, hypertension, arthritis, or severe unexplained pain, are the results of thoughts or attitudes that ultimately result in physical and physiological changes in the body. Lasting imprints upon our subconscious are originally seemingly only innocent thoughts or the relaying of information from one person to another. Thought patterns can, of course, be corrected when the person becomes conscious of the impact of that influence and cooperates in developing a well-balanced, positive personality. Women who seem to have the utmost sex appeal, to be charming and attractive, are not that way merely by chance. The reaction by most people is that if a woman is good-looking, rich, and lucky, therefore, she can act charming because she has all of those things working for her. The exact opposite is really true. I am not suggesting that she deliberately and knowingly took it upon herself to develop her feminine charm to achieve her sexual self-confidence. It was, however, I believe instilled in her early in life, possibly by her mother, someone else very influential, by religious teachings, or by positive attitudinal influences, which assisted her in developing such self-confidence in her life. Becoming a charming person happened due to brain impulses impinging upon the pituitary and the hormonal production, which produced a very feminine lady, who exudes sex appeal, and spiritual, mental, and physical charm.

Medical science has tried for years to identify the illusive factors in women that men find attractive and stimulating. Whether there is an aura that is not measurable medically or not is still unknown. Women, however, who have developed feminine charm, maybe the women who have retained their entire supply of hormones, whether produced naturally or by hormonal replacement therapy. Losing sexual self-confidence negatively affects charm and femininity. The menopausal woman with tragic changes in her life may become one of those unfortunate people who will become a statistic unless replacement therapy comes to the rescue.

Chapter 9: Please Doctor… The Fatal Blindspot

The Breakdown in Communication Between the Sexes

There is a tragic unheard cry of horror stories from American Women who are confused by hormones, aging, and cancer and have yet to have their birthrights restored by hormonal therapy. Many women are subjected to living lives depleted of hormones because most physicians are male and have no personal way of identifying with a woman's experiences. Hot flashes, he believes, may not be imagined in her head, but for other menopausal symptoms, the woman's opinion is devalued. Her symptoms are blamed on the "empty nest" syndrome, or that women often become neurotic in later years of life.

Dr. Wilson, in his book Forever Feminine, describes how his own mother became withdrawn and emotionally difficult to live with during her menopausal period. As a result, Dr. Wilson determined that he would become a physician and try to discover a way to alleviate those physical and emotional symptoms from which his mother was suffering.

A patient recently described the onset of her premenstrual symptoms as beginning slowly, gradually increasing in intensity, until suddenly they would be with the force of a blow. The pain would come in throbs and leave her shuddering. It seemed the soles of her feet ached, and the roots of her hair burned. Deep within was a wrenching twist. She said that she would feel uncoordinated, clumsy, weak, and stumble frequently. Not only would she have double vision, but she felt on the verge of nausea during that time. She reported that she had vile dreams and would awaken with dreadful sweat. This description fits a premenstrual syndrome victim. The effect on spouses, friends, and children is very damaging. "Please, doctor, give me the hormones" is a plea frequently heard from a menopausal woman who should be in the prime of her life. Her skin starts to dry and wrinkle, her strength ebbs, her sex desire wanes, and she fears losing her love partner. She has been told by specialists that it is all in her head or that she has arthritis which is expected with aging. She may end up a tragic victim. This seemingly fatal blind spot in the medical profession can no longer be ignored as nonexistent.

Hormones were discovered and came into use in the forties and fifties. The use of hormones became popular in the sixties, and suddenly, all hormones were suddenly being blamed for cancer due to prescription errors and their use being interrupted. Physicians seemed to become more technician types, becoming less empathetic. They seemed interested in only stopping the pain of arthritis, for example, with toxic medications, rather than correcting the underlying problem of hormonal deficiency and its resultant osteoporosis. The physician seemed to become distant from the patient. Patients became dissatisfied. They seemed insulted by the rudeness and poor results from their expensive visits to the doctor's office. Rudeness, callousness, and aloof attitudes caused the injured party to become enraged and seek legal justice through the court system. The malpractice suit was born. Doctors at first became angry. Later, they became suspicious of every patient who walked through their doors seeking help, whether that patient was inclined to take legal action. So, it is the "state-of-the-art" today that the patient has heard enough about doctors who really don't care about their patients. Every patient has a creeping suspicion in the back of their mind which wonders if the doctor is interested in them personally at all, or if so, how long will their interest last. The doctor who is interviewing the new patient is thinking, "What does this patient really want from me? Will I get into any legal trouble by prescribing specified medications for this patient?" Eighty percent of patients are seeking relief from pain, and eighty percent of the physicians only prescribe pain remedies instead of trying to determine the underlying problem. Therefore, cries for help go unheard because both parties are too busy with suspicious thoughts regarding severe disappointments of each other in their roles. Where there used to be friendship, trust, and celebration over victories from illnesses, now there is distrust, suspicion, and failure.

"Fragmentation Specialists"

In any science that grows at Herculean speeds, as medicine has in the last thirty years, there is a tendency toward specialization. Those who show a preference for certain specialty areas such as the care of allergies, female problems, or arthritis, are free to concentrate on that area only. Generally, a medical student decides during their sophomore year what his specialty area is going to be. He has no feeling usually for that specialty that early in his career. Their decision is based upon prestige or possibly admiration for a doctor or professor in that specialty area. HIS CHOICE may or may not have anything to do with his own personality interests or how he may fit into the specialty area later. Certain specialty areas have a high volume of money income, and that influences many decisions. Frequently, excellent specialists tire or become bored in a few short years after they enter their area of specialty. They cannot wait until Wednesday afternoon comes when they can play golf and try to forget their chosen medical practice in which they feel trapped into performing for eight hours a day. This is tragic, not only for the physician but also for the patient who senses feelings of frustration. The next problem regarding specialization is the fragmentation that takes place when a physician confines his studies and interest to only one area. For example, a patient who had been treated by a urinary specialist came to me. After repeated treatments with antibiotics for urinary tract infections and bladder infections, there was no success. Most urine samples came back sterile or contained no bacteria, yet the symptoms continued. It seems to me that each specialty group of physicians should be taught that a lack of hormones will play havoc with the symptoms they are treating. The lack of hormones causes the membranes to become too thin, dry, and subject to inflammation and infection. Upon analysis, the physician could determine that the bladder lining was so thin that normal spices in the woman's daily diet caused an inflammation, giving her cystitis or bladder infection symptoms. Antibiotics, of course, were not a necessary treatment to correct the underlying cause.

There is also the woman who comes to me with a dry eye condition for which her ophthalmologist prescribed eye drops to which she was severely allergic. Again, a lack of hormones caused dryness of tissues and membranes and the liquids that bathe the eyes. This drying syndrome will continue until this woman's hormones reach their normal level again. Gastroenterologists don't quite seem to relate to the reasons why thinning of the intestinal wall is responsible for symptoms. Orthopedic surgeons do not seem to understand why the bones are getting thin from a lack of hormones. Cardiologists do not seem to realize that a dangerous depletion of hormones causes muscular weakness and the progression of arteriosclerosis in the blood vessels.

The current problems caused by fragmentation are a breakdown of understanding between specialty groups of physicians, which is very serious! Until this dilemma can be solved by either re-educating all currently licensed and practicing physicians or revamping the medical colleges, hoping that a remedy can be reached within the next generation, patients will probably have to exercise their rights. Furthermore, they will have to become more assertive and seek out physicians who will cooperate to bring about the healing process for their own family, friends, and loved ones. To this end, this book is dedicated.

A current newspaper article stated that it is now believed that estrogen should be used as a preventive measure for osteoporosis as well as for cardiovascular disorders. It was further stated that the Scientific Council on Drugs had set forth that taking hormones for lengthy periods increases the chances of cancer from five to fifteen times. It also stated that giving the smallest dose possible for less than two years would not subject the woman to cancer. This is erroneous, dangerous advice. Inadequate therapy can only stop most "hot flashes" and give the patient a false sense of security. The real damage continues! During the sixties, there were many articles written alluding to the book Forever Feminine by Dr. Robert A. Wilson. It stated that hormones were the fountain of youth for women. It would keep them young and wrinkle-free and stop depression, heart attacks, and arthritis. This fell to disrepute in the late sixties and early seventies when the cancer scare began. By reading the book, anyone could understand that Dr. Wilson was correct. Women come to my office every day who did not have any wrinkles even at age seventy, but then some doctor, due to fear or ignorance, took them off of hormones. Three or four years later, they have aged like a biological clock, rushing to catch up with time. When suddenly taken off hormones, these women usually have a heart attack, severe arthritis, or become depressed. They relate that they feel lousy.

Review Dr. Wilson's work, and you discover for yourself that this man was far ahead of his time. Like Dr. Greenblat and Professor Fuller Albright, M.D. of Harvard, he knew that if you are going to fool with Mother Nature, you had better mimic Mother Nature as close as possible. Since females produce the two hormones, estrogen, and progesterone, one should reinstitute these two hormones together, the same as in nature. Doctors who were giving estrogen alone to relieve hot flashes caused estrogen to become out of balance. That fact promoted the proliferation of the endometrial lining of the uterus, leading to cancer. Women who were prone to cancer did get cancer of the uterus by that imbalance of hormones. Now that all of the statistics have been analyzed, it has finally been made public regarding the Twenty-Year Studies. Estrogen and progesterone hormonal replacement therapy proved that women on these hormones had only one-third the death rate from cardiovascular diseases and fatal heart attacks. As Dr. Wilson had alluded, only one-fifth had paralyzing strokes, one-half to one-eighth had cancer of the uterus, breast, and ovaries, and one-tenth had rheumatoid arthritis. Dr. Varon has been indicating this fact for fifteen years. They suffered only one-two hundredths times as much depression and one-two-thousandths as many hip fractures. This report does not even mention the enriched quality of life which is enjoyed by women with proper hormonal levels. Hormonal depletion is no respecter of age. You can call it premature menopause if you wish, but it can happen when a woman is twenty or even to children before they become ten years of age.

Recently, in the course of a conversation over lunch, a lady mentioned that her two sons were both very small for their age. She had taken them to a medical group center in New York City. She had been advised that her sons were hormonally deficient and would need hormones. The two sons were sickly, irritable, thin, and small. Their father had a hormonal deficiency also when he was young. She was afraid to take the boys for hormones because she had heard that hormones caused cancer, so the story continues regarding the misunderstanding of hormones and cancer.

Consumer effort is needed to bring about the education of the lay public, which will affect the health of almost every person living today. Physicians and laypersons must understand the administration of hormones for life enhancement or prime life extension. When a person needs hormones, they should be informed of the minimal risks and reassured that when hormones are given correctly, they rarely could cause cancer. In fact, they have a protective influence against cancer. It is a strange phenomenon in the United States and possibly worldwide that doctors refer patients to various specialists. Yet, very few of those doctors recognize the picture of hormonal depletion. Hormonal depletion can mimic hundreds of other diagnoses, such as rheumatoid arthritis, depression, gastrointestinal problems, and thinning or hyperirritability of membranes of bronchial tubes and sinuses, causing asthmatic and allergic problems. The thinning of the GI tract causes symptoms, including ulcers, diverticulitis, and colitis. Vascular problems result not only in symptoms but in actual death through myocardial infarction. Crippling through paralyzing strokes is often due to a lack of hormones and its resultant severe atherosclerosis. Men and women who have repeated bladder infections, or, more correctly, bladder inflammations, are due to the depletion of hormones.

Urologists, orthopedics, rheumatologists, internists, cardiologists, gastroenterologists, and all other specialists should know exactly the effect of hormones on the area in which they specialize. When treatments, salves, medications, dilatations, etc., do not relieve symptoms, it should be time to look for the real cause of the problem. Physicians are taught in medical school that a host of symptoms from diseases such as syphilis and infectious mono can copy or mimic other diseases. Physicians, in general, are taught to have a high index of suspicion regarding infectious mono or syphilis when they appear to be another disease. The same high degree of suspicion should be true regarding the symptoms which mimic hormonal deficiencies. Physicians should be taught in medical school that symptoms appearing to be arthritis, neuritis, depression, dry skin, skin blemishes and blotches, digestive problems, depression, urinary problems vascular issues, may result from a deficiency of hormones.

These symptoms may indeed be only a disease condition – a deficiency of normal hormones. Three to four percent of people already have adequate hormonal levels either because their glands produce enough hormones or they have kept healthy due to hereditary factors. They may even seem to abuse their body and get away without premature aging. Occasionally, a woman who has a hysterectomy at age thirty will see a doctor who is knowledgeable about hormones. She may stay on tablets and injections, keeping her hormones at a high level for twenty or thirty years. This woman may look younger than her daughter, especially if her daughter has not received hormonal replacement therapy because the doctor of the daughter did not believe in hormones. This is seen over and over again. Patients no longer need to live in fear of getting cancer. In fact, hormonal therapy helps protect against cancer.

The fragmentation of specialists makes it difficult for a woman to determine whether her symptoms are due to something in a specialty area or, the great imitator of all, the lack of hormones. Hormonal deficiency should be considered first as a cause before progressing to other specialty disease areas.

Indications For Hormonal Therapy Individualized Plans

Sample Prescriptions Example

ONE: MENOPAUSE

Natural, After a Hysterectomy or Tubal Ligation

Menopause is the entire complex set of changes brought about in the body due to the depletion of sex hormones regardless of age! In the woman, this is primarily the decline in levels of estrogen and progesterone, the two female sex hormones. Although these changes are usually associated with women in their Sorties and fifties, menopause symptoms can occur at any age. Many cases are due to surgical interruption of the blood supply to the ovaries, most often seen after hysterectomies, tubal ligations, and other pelvic surgeries, even if ovaries are "left in." Any factor that causes a lowering and depletion of estrogen and progesterone causes menopause, which can happen at any age! Inherently low hormone levels may even occur in young women in their teens, causing the "terrible teens," which would be more accurately termed "teenopause!" She experiences changes in the lack of sex desire, the skin begins wrinkling and drying, hair becomes brittle, depression settling in, and other physical symptoms where the body membranes and tissues are becoming thin and fragile, including dry eyes, sinus, intestinal, bladder, and vaginal atrophy. A degeneration of muscles, organs, and bones will be noticed in the early stages.

Menopausal conditions can be triggered by a hysterectomy, thus inducing the above-mentioned symptoms earlier than would have been expected because the surgery itself destroys nearly eighty-five percent of the blood supply to the ovaries. It also occurs in many cases after tubal ligation.

Individualized Treatment

1. Estrogen 1.5 cc, twice a week for three weeks, then twice monthly for six months.
2. Progesterone 1/4 cc s twice a week for three weeks, then twice monthly for six months.
3. Premarin tablets 1.25 mg once or twice daily if tolerated.

Hormone tablets are produced from the urine of mares which contains twenty-three other estrogen substances which are not human in nature. Many women experience adverse reactions when taking these tablets, which are not well absorbed in the gastrointestinal tract. Digestive enzymes often destroy the hormones before they have a chance to be absorbed. Why, then, do we even prescribe tablets? It is a simple matter of economics. The tablets, which forty percent of the women can take, will allow more stability in hormonal level, and the injections can then be spaced further apart. This procedure will prove to be less expensive in the long run. The tablets will make the injections last longer.

Injections last from nine to fourteen days, therefore tablets are needed to also keep the hormonal level stable. When women cannot take the tablets without side reactions, the injections should be given every two weeks or as needed.

Example TWO

MENOPAUSE COMPLICATED BY THINNING OF THE BONES

The type of woman who is most at risk is the small-framed, beautiful woman who has been slim all her life. She is already deficient in bone mass due to her smaller bone structure and dieting throughout her life to remain slim. Her diet has probably been deficient in calcium and protein for years before menopause. This woman has been without hormonal replacement therapy. She now has dangerous thinning of the bones, which may fracture at any time, often causing fatal embolisms where fatty cells from the bone marrow move directly into the bloodstream, reaching the heart or lungs. Non-fatal embolisms lodge in the brain, usually rendering the woman paralyzed and probably senile or psychotic for the rest of her life. Fractured lips that will not heal despite GI surgery cause these women to be placed in nursing homes because they are senile or psychotic and confined to a wheelchair for the rest of their lives. Cases of such confinements caused by hip fractures number eighty-five thousand in the United States alone every year. The human suffering is incalculable, and the cost of spinning medical wheels exceeds three to four billion dollars each year. The results are tragic and unacceptable. Prevention is the only hope.

Individualized Treatment

1. Estrogen 2 cc three times weekly for six to twelve weeks if tolerated.
2. Progesterone 1/2 cc three times weekly for six to twelve weeks if tolerated.
3. Testosterone 1/2 to 1 cc three times a week for six to twelve weeks if tolerated.
4. Premarin 1.25 mg once or twice daily from the first to the twentieth of the month.
5. Provera tablets daily from the fifteenth to the twenty-fifth of each month.
6. Natural B6, or Pyridoxine, 100 to 200 mg. or Lasix 40 mg once or twice a day with each meal with K-Lyte three times a day to control water retention caused by the fast acceleration of hormonal level.
7. If there is a tendency toward congestive heart failure or pulmonary edema, give only one injection every two weeks, as tolerated. Start the tablets after the second or third

injection only if progression is not complicated by any reactions.

8. Add calcium 1500 mg supplement with Vitamin D 400-800 mg. per day.
9. Any activity, walking, swimming, or even simply rocking in a rocking chair as appropriate.
10. Exposure to ultraviolet rays or sunshine for at least ten
11. minutes three or more days per week.
12. A nutritionally balanced diet, including appropriate amounts of complex carbohydrates and protein. See the sample diet in the appendix.
13. Eliminate coffee with caffeine, alcohol, and nicotine, as these will cause further destruction of the bone mass.
14. Safety — Hallways and stairs should be well-lit. Avoid throw rugs and slippery surfaces.

Example Three

Premenstrual Tension

A clinical diagnosis indicates a woman who becomes impossible to live with five to fifteen days before her period each month and when certain symptoms occur at definite intervals during the cycle. These women unknowingly forfeit anywhere from twenty-five to fifty percent of their normal, happy life. Premenstrual tension is definitely an endocrine disorder caused by a hormonal deficiency of progesterone. The clinical diagnosis is confirmed by having the patient chart the monthly cycle, write in the days, foods craved, foods eaten, bloating if present, emotions if unusual, aggressive hostility, and lack of sexual satisfaction. Chart the cycle of the periods and determine any relationship between symptoms and the cycle. If she seems to feel calm by the first or second day of her period, she is dealing with premenstrual tension.

Individualized Treatment

1. Progesterone 1/2 cc injections weekly.
2. Dietary controls.
3. Anti-depressants, if needed, for assistance with depression. Negative thinking and depression must be corrected to change signals for utilization by the body of the hormones received.
4. Counseling is encouraged.
5. Diuretics in the form of natural Paradoxine, which is B6, 100-200 mg with each meal. If a stronger

chemical drug diuretic is used, it must be supplemented with Potassium Chloride ten percent liquid or effervescent tablet called K-Lyte. One tablespoon of liquid or one effervescent tablet per each given diuretic tablet.

6. Omit salty foods.
7. Omit sweets, alcohol, and caffeine (which may be craved at this time).

Example Four

Pregnancy-Related Progesterone Deficiency

A woman who has had repeated miscarriages or had a difficult time with her last pregnancy, retaining liquids, becoming hypertense, etc., or a woman who has had a toxemic pregnancy. She should not be pregnant at the time of hormonal replacement therapy.

Individualized Treatment

1. Progesterone 200-400 mg or '/2 CC three times weekly or by injection, a vaginal suppository of 200-400 mg. progesterone during the last two weeks of the monthly cycle. Suppositories are not acceptable to most women; however, if she cannot get injections three times a week for two weeks prior to each period, then substitute injections with vaginal suppositories before retiring at night. The medication in the suppositories must be absorbed through the vaginal walls into the bloodstream. Absorption will occur within six to eight hours

and should not be interfered with by tampons.

2. Estrogen 1.5 cc three times weekly during the first part of the month.
3. Testosterone 1/2 cc once each month mid-cycle if indicated by lack of enthusiasm or sex drive.

Example Five

THYROID DEFICIENCY "Twin Sisters"

A woman lived her early years in an area where the iodine had leached out of the soil. Families from around the Great Lakes area or the State of Washington have for generations tend- ed to have hyperthyroidism, overactivity, or hypothyroidism, with a goiter condition.

Individualized Treatment

1. Psydomil 25 micrograms or T-Thyroxin which is called Synthroid 0.2 mg (Synthroid has a longer shelf life than Armus thyroid). 8 PM temperatures (halfway between supper and bedtime) should be taken to determine if the body temperature registers below 98.4, remembering that 98.6 is normal. If the 8 PM temperature is 98.2 or below, prescribe more than one tablet with breakfast each day. Usually one tablet with breakfast, 1/2 tablet with lunch and 1/2 at dinner, for the first ten days of each month will bring about changes in three to four months.
2. Estrogen 1.5 cc every two weeks.

3. Progesterone '/2 cc every two weeks will be needed in a cycling pattern for six months to one year. (Birth control pills or the hormone tablet called Premarin, 1.25 mg, and Provera in a cyclic manner could be prescribed in place of injections. Premarin would be taken from the first to the twentieth day of her calendar month.)

Example Six

Teen-O-Pause

Teen-o-pause is premature menopause occurring during the teenage years in girls. A girl with teen-o-pause was formerly charming, self-satisfied and obedient, and suddenly at the age of about thirteen, becomes rebellious, unkept, and unable to concentrate in school. Even if she was highly intelligent and received good grades, she changes to become restless or becomes a cause of disruption in the classroom. Frequently, this is the girl who becomes so distraught or antagonistic that she appears to be in competition with her mother for the love and approval of her father. This competition often brings about arguments, fights, and even broken marriages. Most commonly, this girl is simply confused, depressed, bloated, having difficulties with her emotions, and adjusting to changes within her physical body, perhaps even her menstrual cycle.

Individualized Treatment

1. Birth control pills should be rotated for six to nine months in

sequential amounts of estrogen and progesterone. It may not be enough to make up the difference.

2. Injections of estrogen: Progesterone. Low dosage for first fourteen days. Progesterone, high dose estrogen, moderate dose the last fourteen days of her cycle.

After sixty to ninety days, a saturation effect will take place, and this girl will appear "caught up" with her hormone level. She will again become the happy, complacent, and charming young lady she was before her hormone depletion.

Chapter 10: The Promise of Realistic Expectations

Can a young lifestyle be rejuvenated? Can correct hormonal replacement rejuvenate the brain, tissues, and bones?

After a period of destruction and disintegration due to a lack of hormonal levels, there can only be a realistic expectation as to the amount of recovery. Dowager's hump and pain in the spine may respond to hormonal replacement therapy, calcium, and vitamin D therapy, and there may be a reduction or elimination of pain. However, neuritis caused by the bones compressing on the nerves cannot be reversed. Disfiguring from the Dowager's hump is also relatively permanent. A first heart attack is often fatal, and senility, caused by fatty cells from a fractured hip lodging in the brain, are irreversible tragedies. Few women ever return to their former mental, physical, or emotional levels after fracturing a hip. Prevention is the key, and hormones should be used for preventative purposes to avoid the damages now known to be caused by hormone deficiencies. Hormonal replacement therapy will bring about positive changes if it is started before serious and permanent damage. Even though the level of hormones can be measured, it is still impossible to determine the quality of hormones present in a woman's body. When there is any doubt, it would be better to give a trial of hormonal replacement therapy than to suffer the consequences of a hormone deficiency. Prevention is the key to a rich, fulfilled life well into the age of ninety.

"The Promise"

Ten Commandments Of Hormones Which Will Bring Promise For A Richer Life

1. Proper hormonal balance will increase the quality and length of life.
2. Hormonal therapy will reduce the risk of cancer of the uterus, breast, and ovaries by over fifty percent.
3. Hormonal replacement therapy will diminish osteoporosis, low back pain, and hip fractures by over ninety percent.
4. Hormonal replacement therapy will relieve and eliminate pain in ninety-two percent of rheumatoid arthritis cases.
5. Proper hormonal balance will assist in retaining youthful skin, hair, figure, and muscle tone.
6. Adequate hormonal levels will increase energy, memory, sleep, libido, and vivaciousness.
7. Hormonal replacement therapy will eliminate most symptoms of depression, irritability, premenstrual tension, and menopause.
8. Adequate hormonal balance will lessen intestinal and urinary problems.
9. Proper hormonal levels will eliminate hot flashes, sweating, headaches, fatigue, depression, and anxiety.
10. Naturally, high hormonal levels or hormonal replacement therapy will increase a sense of well-being, sensuality, and emotional welfare.

Chapter 11: The Future With Hormones

This chapter is virtually unwritten.

Exciting changes are bringing miraculous possibilities to the forefront. Medical information and updates will be given through seminars, clinics, and monthly newsletters.

Contact us for updates and further information.

Made in the USA
Monee, IL
04 November 2026

67387292R00075